CW00455598

LOCAL AUTHORITY HEAI
ENFORCEMEI

Local Authority Health & Safety Enforcement

Christopher N. Penn, M.C.I.E.H.

*Former Director of Environmental Health and Consumer
Services, Walsall Metropolitan Borough Council*

Shaw & Sons

Published by
Shaw & Sons Limited
Shaway House
21 Bourne Park
Bourne Road
Crayford
Kent DA1 4BZ

www.shaws.co.uk

© Shaw & Sons Limited 2005

ISBN 0 7219 1650 3

Published January 2005

A CIP catalogue record for this book is available from the
British Library

*No part of this publication may be reproduced or transmitted in
any form without the written permission of the copyright holder*

Printed in Great Britain by
Bell & Bain Limited, Glasgow

SUMMARY OF CONTENTS

CONTENTS

TABLE OF STATUTES

TABLE OF STATUTORY INSTRUMENTS

TABLE OF CASES

INTRODUCTION

Despite a plethora of health and safety legislation, particularly over the last thirty years, workplace accidents and ill-health continue at a significant level. Although most deaths occur in the construction industries, much of which are policed by the Health and Safety Executive, high accident levels still occur in the local authority enforced employment sector. This is despite increasing legislation and ostensibly greater knowledge and awareness of health and safety requirements by employers. There are also new challenges to be addressed as the service sector expands and new activities and risks evolve, e.g. a greater number of residential homes and the impact of stress associated with certain workplace activities.

The government's and the Health and Safety Commission's dissatisfaction with the control of workplace health and safety was reflected in its *Revitalising Health and Safety* Strategy Statement published in 2000. This was followed by a Strategic Plan for 2001-2004 containing national targets for the health and safety system in this country. Those targets are set for both the Health and Safety Executive and for local authorities as the two principal enforcement arms of health and safety legislation.

The Strategic Plan sets out ways in which the Health and Safety Commission proposes to achieve the targets that have been set. These include working with various organisations, including local authorities, in a co-ordinated way. The Commission's contribution to achieving the targets is to include taking action in priority areas; ensuring an effective regulatory regime in the major hazard sectors; securing compliance with the law; and meeting the mandate given to it by government.

Despite this commitment to joint working, in July 2003 the first signs of a possible split in working relations started to appear, with a question mark being raised over the performance of local authorities in enforcing health and safety at work legislation. The Head of the Health and Safety Executive Local Authority Unit questioned whether health and safety enforcement should be removed from local authorities, possibly by 2010. The issue was raised because it was said that the current split of functions was not working. There has been a relentless rise in the number of service sector activities enforced by local authorities, without a corresponding rise in the resources to enforce the law. It was suggested that local authorities are "incapable of enforcing health and safety in multi-sited companies". There was also a threat by the Chairman of the Health and Safety Commission to "name and shame" allegedly under-performing authorities. This would have done nothing to improve the morale of hard-pressed local authority enforcement staff. Not only that, the Health and Safety Executive is hardly in a position to criticise when its own inspection rates have fallen over the years. There has also been criticism by the legal profession of certain aspects of the Executive's enforcement role. The clear picture is that both of the primary enforcing agencies could do better!

An increasing range of statutory responsibilities imposed on local authorities clearly requires a review of priorities and methods of working and, although criticism has been made of local authorities' performance, the Chairman of the Health and Safety Commission has also recognised that "they are a key part of delivery" (of health and safety) and also commented that "I don't think that the Commission, the Executive or local authorities have fully reflected the changes in the workplace". He may well be right.

In any event, the Health and Safety Executive Local Authority Unit has begun a series of audits to measure compliance with the Health and Safety Commission's section 18 guidance. It has also worked closely with the Association of London Environmental Health Managers which has conducted a comprehensive inter-authority audit of health and safety management. The audit covered almost all of the London boroughs. It identified a number of shortcomings and challenges which the authorities are now actively tackling. It also identified a number of examples of good practice. The willingness of those authorities to share their experience, findings and approach to the challenges identified shows a positive commitment to enforcement in the local authority sector. This commitment is also evident from many authorities across the country who have willingly provided information and examples of their positive approach to enforcement, at a time of increasing demand for services and when their resources are stretched.

Despite questions about the continuing enforcement of health and safety legislation by local authorities, the Commission does in fact recognise the important contribution that local authorities can continue to play in health and safety enforcement. In its *A strategy for workplace health and safety in Great Britain to 2010 and beyond*[1] it recognises the fact that both HSE and local authority resources are limited. However, it also identifies the need to target those resources to where they will do most good; to involve businesses more in improving health and safety; to develop new ways of working which involve employers and employees more; to do more to address emerging work-related health issues such as stress; and to use the best practices that have been developed.

Whilst the Commission cannot see any "lasting logic" to the current division of enforcement responsibility between the Executive and local authorities, it clearly sees benefit in a closer partnership involving local joint planning; joint decision making; agreed targets; the development of best practices; sharing intelligence, training and expertise; and reviewing the existing liaison arrangements. It has also put on hold a review of the existing enforcing authority regulations pending a wider review of the health and safety system. A review of guidance and local authority circulars is also in progress.

The Health and Safety Executive is also in the process of developing regional partnerships and has appointed seven regional managers. It will also be appointing Environmental Health Officers, as the principal local authority enforcement staff, to each of those partnerships. These teams will look at particular issues, e.g. communications, data capture and collection, the enforcing authority regulations and the allocation of responsibility. The Executive will also extend its support for local authorities through its specialist officers and laboratories.

---

[1] February 2004, HSC.

The Health and Safety Executive believes that the profile of health and safety at work needs to be raised if progress is to be made. In this respect it is targeting three specific areas:

- Elected members.

- Senior managers.

- Health and safety practitioners.

It recognises that elected members and senior managers may restrict what the practitioners are able to do, but it may be hard work changing the attitudes of those members and managers who have consistently failed to recognise the importance of protecting the health and safety needs of those employed in local authority controlled activities.

Whilst greater partnership involvement appears fine, it does not address the problem of those authorities which continually fail to meet the mandatory requirements of section 18 of the Health and Safety at Work etc. Act 1974. Section 45 of the Act provides a way of removing the enforcement powers from individual authorities but it has never been, nor is likely to be, used due to the lengthy processes involved. Maybe a change in the law is needed to produce a speedier way of dealing with such authorities.

Despite, or as a consequence of, the difficulties mentioned above, the Health and Safety Commission, the Health and Safety Executive and the local authority associations have produced a "high level" statement of intent which was unveiled at the Chartered Institute of Environmental Health conference in September 2004. Its seven key principles are intended to ensure closer working relationships and it is to be hoped that the new approach to joint working will help to improve workplace health and safety standards.

The enforcement allocation regulations are likely to undergo significant change, possibly with a view to more flexibility in the allocation of activities between the Executive and local authorities. This could be useful, especially if joint regional plans for enforcement were also produced identifying the joint resource (of the HSE and local authorities) available to tackle the highest priority issues and risks on an agreed basis, and disregarding some of the allocation criteria currently contained in the enforcing authority regulations.

Despite the changes being considered, a number of clear messages emerge:

- The recognition by the Commission of the local authority contribution to its priority targets.

- There is a need to change the existing enforcement arrangements to improve the control of workplace health, safety and welfare.

- Unless all local authorities are prepared to recognise the importance of health and safety enforcement, change will be imposed on them.

- Those experienced environmental health practitioners working in commerce and industry can influence the attitude and approach of their local authority colleagues by making them aware of the issues in private sector activities which influence their own approach to health and safety.

- There has to be a better and more productive relationship between the Health and Safety Executive and local authorities if significant progress is to be made in the enforcement of health and safety at work.

If local authorities are to maintain their role as key enforcers of health and safety legislation, they must take the opportunity they now have to demonstrate their ability to respond positively to their responsibilities. They must identify their service requirements and manage their priorities effectively and efficiently. This may mean in some cases completely reassessing their approach, stopping putting resources into low risk premises and activities, and identifying those areas where they can really make a difference. There are many local authorities taking an innovative, pro-active and well managed approach to health and safety. Those that currently struggle could be well rewarded by seeking their advice and expertise.

The Health and Safety Commission rightly places emphasis on certain priority programmes to deal with the major causes of accidents and ill-health. However, some of those causes, such as slips, trips and falls, have been a problem for years. It raises the fundamental question as to why the Commission and local authorities have only in recent times recognised and started to concentrate more on those issues. It may of course reflect the fact that both enforcing organisations have been over-stretched for several years. If so, perhaps there is a need to look at other alternatives for addressing the resource problems, e.g. if neither organisation has the resources to carry out its role to a satisfactory level, maybe some central funding could be allocated to encourage retired health and safety practitioners to fill the resource gap where there is a significant need.

Perhaps there are other priority issues that also need to be addressed, for example:

• Tackling the problem of children under 16 years of age who may not be getting adequate health and safety protection.[2]

• Addressing the allegedly high levels of personal abuse in care homes for the elderly, although such matters are also likely to involve the police.

• Examining the health and safety risks to immigrants, many of whom disappear into the black economy or find work in the hotel and catering industry. Many of these people will lack knowledge or understanding of the English language and their employers may have little incentive to provide them with health and safety requirements.

• Taking on those enforcing authorities who are not, for whatever reason, fulfilling their obligations to enforce health and safety law, using powers under section 45 of the Health and Safety at Work etc. Act 1974 if necessary.

• Tackling some of the main employer failings which result in accidents, e.g. inadequate instructions and training, poor risk assessments and unsafe systems of work.

---

[2]   See *Environmental Health Journal*, March 2004.

It may need to be accepted that many small to medium sized employers will only fully embrace health and safety if they can see that it will improve their profitability. In many cases it will not. In such cases a different approach may be needed. A pilot scheme which involves putting in a combination of expertise appropriate to a business's needs, e.g. health and safety, food hygiene, waste disposal and pollution control, could result in an improvement in the management of one or all of the selected areas and provide financial benefits to the employer's overall performance. Linked to other relevant business skills providers, employers otherwise reluctant to embrace health and safety may achieve worthwhile benefits.

Notwithstanding the efforts of the enforcing authorities and the increasing amount of health and safety legislation, it is clear that there are many employers, including national employers, who still pay too little attention to the basic requirements of the law and have very poor safety records. Perhaps now is the time to consider a "naming and shaming" policy, the introduction of certification of workplace activity and/or the public display of the findings of health, safety and welfare inspections.

One thing is clear, unless greater priority overall is given to health and safety enforcement, there will be more accidents waiting to happen and increasing ill-health in the country's workforce. If this happens, employers and enforcers will have failed the working population.

This book does not seek to argue about who should enforce different elements of health and safety legislation: that is a matter for the Health and Safety Commission and government. However, as a former local government officer I remain committed to the principle of local enforcement by dedicated and professional officers who are readily accessible and accountable. Accordingly, I have tried to write a book which:

• Sets out the law as it relates to the duties and responsibilities of local authorities in enforcing health and safety requirements.

• Contains lots of case law references to demonstrate the courts' and employment tribunals' attitude to offences.

● Gives examples of good and not so good enforcement practices.

Where I have included examples of shortcomings in enforcement practices, it is simply to help readers consider if they have the same problems. Where I have referred to good practices of some of the best councils, it is in the hope that their experience and expertise may stimulate others to review their methods of work and achieve standards of performance worthy of praise from the Commission.

The breadth and depth of health and safety legislation and practice is daunting. It would not be possible to cover every aspect of it in a book of this length. What I have tried to produce is something which acts as a guide to health and safety practitioners in local government by dealing with the main elements of enforcement they are likely to encounter. I recognise that I may not have dealt with matters in the depth that some may prefer but I hope that the numerous references I have included will direct the reader to other sources of useful information.

I have attempted to state the law as it existed up to 1st September 2004.

Chris Penn

ACKNOWLEDGEMENTS

The idea for this book arose from someone else's view that there was no publication dealing specifically with the enforcement of health and safety at work legislation by local authorities. Having started work on this project I soon realised why! The breadth and complexity of health and safety law and practice is such that it is impossible to put it all into a single book whilst fulfilling the expectations of all those who might wish to read it. Not only that, it is difficult to keep up with the pace of change.

As a result I have tried to produce what I hope is an informative guide to the law and enforcement practice on health and safety at work as it applies to local authorities. In doing so I have had regard to what various practitioners have suggested would be a useful approach. I recognise that most people will want to explore certain matters in more detail. I have dealt with this by including a lot of case law and employment tribunal references, together with reference to circulars and some specific internet websites which contain valuable information. I have also been helped by some experienced and progressive health and safety practitioners in a variety of local authorities across the country. They have readily provided examples of the ways in which they deal with health and safety issues. Most of them are under pressure because of shortages of staff. Their willingness to help in those circumstances is very much appreciated, as is that of members of the Chartered Institute of Environmental Health and Salford University. They are too numerous to mention individually. My wife Sandra has, as usual, been a constant source of encouragement and support.

I mention the following individuals in particular as they have not only provided valuable information and advice, but also offered invaluable comments and suggestions on areas of the text relating to their particular expertise. I remain personally responsible for any errors or omissions.

Josie Brown, Salford University.

Allan Davies, Head of Local Authority Unit, Health and Safety Executive.

Adrian Levett, Head of Trading Standards, Warwickshire County Council.

John Pointing, Barrister, Field Court Chambers, Gray's Inn, London.

Phil Preece, Health and Safety Manager, Regulatory Services, Birmingham City Council.

C.N.P.

Chapter 1

COMMON LAW LIABILITY

RELATIONSHIP TO STATUTORY LAW

Although this book deals primarily with the statutory role of local authorities, decisions at common law impact on the way decisions are taken when enforcing health and safety legislation. Common law precedents also often form the basis of statutory law. It is therefore relevant to make some reference to the key elements of common law as they relate to health and safety.

It is in fact quite possible for a single health and safety incident or accident to give rise to several courses of legal action. For example, an employee may be injured working on a new piece of equipment. If that equipment was supplied in a faulty condition there may be a case for suing the supplier and the employer in negligence. There may also be a breach of the employee's contract of employment. Under the primary statutory legislation, the Health and Safety at Work etc. Act 1974, there may be breaches of the law relating to the supply of the offending article; the systems of work and training involved; and in addition there may also be breaches of the law relating to product liability enforceable by trading standards authorities.

In respect of health and safety, common law action involves the right of an employee to sue an employer for damages in respect of death, personal injury or disease.

LIABILITY OF AN EMPLOYER

The liability of employers at common law is part of the general law of negligence. A claimant has to show that:

(a) the defendant owed him a duty of care;

(b) the defendant was in breach of that duty;

(c) as a result of that breach the claimant suffered damage.

In practice, when the issue of liability arises, the following questions will inevitably arise:

1. Did the defendant owe the claimant the duty in respect of the act or omission complained of?

2. If so, was the standard of care shown by the defendant less than that required of him?

3. Can the claimant prove that there was negligence by the defendant, and that the negligence was the cause of the damage?

4. Are there any special defences available which would defeat the claim, either in whole or in part?

Where the employer is the Crown,[1] it is basically under the same liability to its employees as any other employer,[2] although the Crown is not bound by a statutory duty unless the statute expressly states or implies so. An employer who is liable[3] to an employee for damages will, under the general law,[4] have a claim for partial or complete reimbursement against any other person who may be liable for the same damage, e.g. a negligent employee for whom the employer has been held vicariously liable, a manufacturer of defective equipment, a temporary employer to whom the general employer has entrusted the safety of his own employee,[5] or the owner or occupier of a building where the employee is sent to work and is injured by a defect in the building.[6] Damages claims may survive the death of the claimant and pass to his executors or administrators, even if the claim had not been established at the time of death.[7]

---

[1] e.g. government departments, HM Forces, and a number of other bodies such as the Forestry Commission, Medical Research Council, etc.
[2] Crown Proceedings Act 1947, s.2(1)(b).
[3] The employer will not be liable at all if the other parties against whom action is taken have paid the employee a sum in full satisfaction of his claim: *Jameson v. Central Electricity Generating Board, The Times,* December 17, 1998, HL.
[4] Civil Liability Contribution Act 1978.
[5] *Nelhams v. Sandells Maintenance Ltd., The Times,* June 15, 1995, CA.
[6] *Andrews v. Initial Cleaning Services and Another, The Times,* August 18, 1999, CA (75% liability was apportioned to the employer, 25% to the building owner).
[7] *Ronex Properties Ltd. v. John Laing Ltd., The Times,* July 28, 1982, CA.

Where an employee has contracted an industrial disease, and it might have been contracted at any time whilst the employee was working for a succession of employers, the inability of the employee to demonstrate in which employer's service it occurred does not mean that none of the employers will be held liable. In fact, any employer who materially increased the risk of the employee contracting the disease will be liable.[8] A well-established example of this principle is in the case of asbestos-related mesothelioma.

If the defendant employer is a company and is defunct, it can in appropriate circumstances be restored, by the court, to the Companies Register to enable a personal injuries action to be brought against the company.[9]

As well as a claim against his employer for damages, an employee may also claim against his employer for breach of statutory duty and against the state under the industrial injuries and diseases national insurance scheme.[10]

DUTY OF CARE

General scope of the duty

The ruling principle is that the employer must take *reasonable care* for the safety of his employees and other persons affected by his operations. The common law requirements are that employers provide and maintain:[11]

(a) a safe place of work with safe means of access and egress;

(b) safe appliances and equipment and plant for doing the work;

---

[8]   *Fairchild v. Glenhaven Funeral Services Ltd.* (and other conjoined cases), *The Times,* June 21, 2002, HL. As to costs payable in this type of multi-party action, see *Africa and Others v. Cape plc, The Times,* January 14, 2002, CA.

[9]   *Re Workvale* (1991) B.C.C. 109 (insurers of company party to application to restore); *Re Philip Powis Ltd., The Times,* March 6, 1998, CA (declaration that company dissolution void to enable claim for damages for personal injuries to be made, even though *prima facie* out of time but possibly extendable under the Limitation Act 1980, s.33).

[10] Embodied in the Social Security Contributions and Benefits Act 1992 and regulations made thereunder, principally the Social Security (Industrial Injuries) (Prescribed Diseases) Regulations 1985, S.I. 1985 No. 967.

[11] *Wilsons and Clyde Coal Co. Ltd. v. English* [1938] A.C. 57.

(c)   a safe system of work and effective supervision; and

(d)   competent and safety conscious employees.

In general, the duty of an employer in cases of negligence only extends to protecting the employee against personal injury. However, this may include psychiatric injury, such as a nervous breakdown caused by stress or overwork.[12] It does not include protecting the personal property of an employee, but would extend to a claim for clothing and any other personal property damaged in the course of personal injury.[13] If an employer orders an action which may imperil the property of an employee known to be in a certain place, the employer may be liable under the general law of negligence.[14] The employer's duty *may not necessarily* extend to his relatives, for example sending an employee home with asbestos dust impregnated clothing, thus giving asbestosis to the employee's wife.[15] Neither does it extend to warning or insuring the employee against economic loss likely to be incurred in the employment.[16]

The duty to take reasonable care to protect an employee against personal injury does not extend to the case where the employee

---

[12]   *Walker v. Northumberland County Council* [1995] I.R.L.R. 35 (manager's nervous breakdown). The subject of employer's liability for mental or physical injury caused by stress at work was reviewed in *Hatton v. Sutherland, The Times,* February 12, 2002, CA. It was held that the employer was liable only if he could see that the particular employee could suffer injury by stress and failed to take reasonable steps to prevent it. Proof of occupational stress was not, of itself, enough. See also *Pratley v. Surrey CC* [2003] I.R.L.R. 794, CA (forseeability of future risk of psychiatric illness not enough to impose liability for immediate mental collapse). An employer will also be vicariously liable for one employee's failure to assist another employee (if under a duty of care to do so) who has complained to him of work stress problems: *Barber v. Somerset CC* [2004] UKHL 13; *The Times,* April 5, 2004, HL; reversing the Court of Appeal in the case: [2002] I.C.R. 13.

[13]   *Deyong and Shenburn* [1946] K.B. 227, CA; *Edwards v. West Herts Group Hospital Management Committee* [1957] 1 W.L.R. 415, CA.

[14]   *ibid.,* per Tucker J.

[15]   *Gunn v. Wallsend Slipway and Engineering Co. Ltd., The Times,* January 23, 1989 (no claim in respect of wife's consequent death). *Cf. Hewett v. Alf Brown's Transport Ltd.* [1992] I.C.R. 530, CA (no duty to wife poisoned by lead in husband's clothing, as no breach to husband at common law or under the Control of Lead at Work Regulations 1980, S.I. 1980 No. 1248).

[16]   *Reid v. Rush and Tompkins plc* [1989] 3 All E.R. 228, CA (employee injured abroad by the negligence of an untraceable third party driver). If the risk occurs in Great Britain, the employer must insure under the Employers' Liability (Compulsory Insurance) Act 1969.

goes off on his own to do work which is neither required nor expected.[17] However, if an employee does work which, although not part of his duties, might nevertheless be reasonably expected of him, a duty of care will be owed to him.[18]

The place where the duty of care is owed will normally be the premises of the employer and any outside place of employment, including transit to and from the main place of employment. The duty of care will not generally extend beyond those places.

Where an employee suffers from a known abnormal sensitivity, e.g. dermatitis from grease, and if he wishes, despite his knowledge of the risk, to continue at a job exposing him to that risk, the employer may be under no duty to refuse to employ him or to dismiss him from a job involving such risks, providing that the employer ensures that the employee has a full understanding of the risks involved.[19] This principle will not apply if the risk is considerable and the employee has full knowledge of it.[20]

It would appear that an employer also owes a duty of care to take reasonable precautions to protect an employee from physical violence whilst performing his duties, and this would appear to be the case whether the violence is committed by a fellow employee,[21] or by a stranger.[22] However, the employer would only be regarded as negligent if he did not take steps to eliminate a risk which he knew or ought to have known was a real risk and not merely a possibility which would never influence the mind of a reasonable man.[23]

---

[17] *Hayden v. Mersey Docks and Harbour Board* [1956] 2 Lloyd's Rep. 497.

[18] *Tomlinson v. Ellerman's Wilson* [1959] 1 Lloyd's Rep. 497.

[19] *Withers v. Perry Chain Co.* [1961] 1 W.L.R. 1314, CA; *Kossinski v. Chrysler United Kingdom Ltd.* (1974) 15 K.I.R. 225, CA.

[20] *Cork v. Kirby McLean* [1952] 2 All E.R. 402, CA (employment of a known epileptic on scaffolding); *Coxall v. Goodyear Great Britain Ltd., The Times,* August 5, 2002, CA (employer should have removed employee known to have predisposition to asthma from work safe to an ordinary employee but unsafe to *that* employee).

[21] *Smith v. Ocean S.S. Co.* [1954] 2 Lloyd's Rep. 482.

[22] *Houghton v. Hackney Corporation* (1961) 3 K.I.R. 615.

[23] *Per* Lord Reid in *The Wagon Mound (No.2)* [1967] 1 A.C. 617 at 642, HL, cited by Lord Denning M.R. in *Charlton v. Forrest Printing Ink Co. Ltd.* [1980] I.R.L.R. 331 at 333, CA, in a case in which, on the facts, employers where held not negligent in sending an employee who was robbed and nearly blinded, to collect wages of £1,500, instead of employing a security firm. Whether this decision would continue to be upheld in today's even more violent society remains to be seen.

The risk of repetitive strain injury to those undertaking regular typing work is one likely to be frequently encountered by local authority inspectors. It has been held that those employees who are likely to do a great deal of typing need to be told that they must take breaks from prolonged spells of typing.[24]

STANDARD OF CARE

The standard of care generally expected of an employer is that of an ordinarily prudent employer.[25] The standard of care is not to be increased by virtue of the fact that employers usually insure against accidents and injury.[26] The standard of care must not be set so high that it becomes indistinguishable from the absolute statutory obligations of the employer.[27]

The standard of care required in law is as follows.

Foreseeing the existence of the risk

If the work contains an element of danger, the employer must foresee the possibility of injury occurring even if it involves an element of fault by the employee and that fault could not be forecast precisely.[28] However, for the risk to be reasonably foreseeable, the accident must be one which could or would be reasonably foreseeable "in the ordinary course", as opposed to an event which would be wholly exceptional, or which no reasonable employer could be expected to anticipate.

Assessing the magnitude of the risk

The seriousness of the injury at risk and the likelihood of it actually

---

[24] *Pickford v. Imperial Chemical Industries plc* [1996] I.R.L.R. 622, CA.

[25] *Paris v. Stepney Borough Council* [1951] A.C. 367, HL, *per* Lord Oaksey; *Smith v. Austin Lifts* [1959] 1 W.L.R. 100, HL, *per* Lord Morton. See the exposition of the duty by Swanwick J. in *Stokes v. Guest Keen and Nettlefold (Bolts and Nuts) Ltd.* [1968] 1 W.L.R. 1776; *Pearce v. Round Oak Steel Works Ltd.* [1969] 1 W.L.R. 595; and *Flynn v. Vange Scaffolding and Engineering Co., The Times,* March 26, 1987, CA.

[26] *Davie v. New Merton Board Mills* [1959] A.C. 604, HL.

[27] *Latimer v. A.E.C.* [1953] A.C. 643, HL. *Cf. Thomas v. General Motors-Holdens* (1988) 49 S.A.S.R. 11.

[28] *Thurogood v. Van Den Berghs and Jurgens* [1951] 2 K.B. 537; *sub nom. Thorogood v. Van Den Berghs and Jurgens* [1951] 1 All E.R. 682, CA.

occurring must both be taken into account.[29] It may also be necessary to consider the consequences of *not* assuming a risk. It must also be considered that the standard of care owed to someone with a known disability is greater than that owed to someone without a disability. In many activities some risk may well be unavoidable, and in such cases as long as the employer takes reasonable care to conduct the activity in a reasonably safe way, he is undertaking his obligation.[30] It has been said that "one must always guard against making industry impossible and against slowing down production by setting unduly high standards or by placing wholly unreasonable requirements on the makers and owners of machinery of this kind."[31] Whether such a view would be taken in the case of action under the Health and Safety at Work etc. Act 1974 to deal with a high risk situation would depend on the precise circumstances.

Nevertheless, although in common law the fact that an activity involves some risk *does not necessarily mean that the employer will always be liable for any accident that occurs*, the standard of care is a high one.

Devising reasonable precautions

In considering whether some precautions should be taken against a foreseeable risk, the employer has a duty to consider on the one hand the magnitude of the risk, the likelihood of an accident occurring and the possible seriousness of the consequences if an accident does happen and, on the other hand, the difficulty and expense and any other disadvantage of taking the precautions.[32] In deciding whether certain precautions ought to be taken, it will often be important to see what other employers do in similar cases, although what others do will not always be conclusive.

---

[29] In *The Wagon Mound (No.2)* [1967] 1 A.C. 617, PC, Lord Reid said that a "person must be regarded as negligent if he does not take steps to eliminate a risk which he knows or ought to know is a real risk and not a mere possibility which would never influence the mind of a reasonable man."

[30] *Newman v. Harland & Wolff* [1953] 1 Lloyd's Rep. 114.

[31] *Jones v. Richards* [1955] 1 W.L.R. 444.

[32] *Morris v. West Hartlepool Steam Navigation Co.* [1956] A.C. 552, *per* Lord Reid; *Gawtry v. Waltons Wharfingers & Storage Ltd.* [1971] 2 Lloyd's Rep. 489, CA. Where the possibility of injury is remote, and no better practical method of doing the particular job is known, no liability will arise in common law: *Hindle v. Joseph Porritt & Sons Ltd.* [1970] 1 All E.R. 1142.

Abnormal susceptibility of employee

If an employer exercises a sufficient degree of care to provide reasonable protection for the normal employee and does not know nor ought to have known about the abnormal susceptibility of an employee, he will not be liable for injury arising from that abnormality. However, if he knew or ought to have known of the abnormal insensitivity, he will usually be under a duty to take extra precautions in relation to that person. Presumably this implies that the employer must ask relevant questions prior to the employment of individuals, and also monitor employees to determine if any such sensitivity arises. He may have this responsibility under statutory law in any event. Equally, an employee knowingly suffering from some abnormal susceptibility which increases his risk of injury is under a duty to disclose this to his employer.[33] Where an employee suffering from a known abnormal susceptibility wishes to carry on with his job despite a full understanding of the risks involved, then he may do so.[34] It is questionable whether the employer or the employee would get away with this arrangement under statutory law, and under common law the principle would not apply where there was a risk of death.

Instructions and supervision

In the case of skilled men, instructions will *not normally* be required where the work is within their competence. However, this is not without some doubt.[35] In most cases it is reasonable to assume that instructions will be required and in such cases the instructions must be positive.[36]

If an employer allows an obviously unsafe practice to develop and continue and does nothing to stop it, he will be at fault.[37] The

33   *Cork v. Kirby Maclean* [1952] W.N. 399, CA.
34   *Withers v. Perry Chain Co.* [1961] 1 W.L.R. 1314, CA; *Kossinski v. Chrysler United Kingdom Ltd.* (1974) 15 K.I.R. 225, CA. *Cf. Richardson v. Telehoist* [1981] C.L. 299 (unreported), CA.
35   In the case of *Winter v. Cardiff R.D.C.* [1950] W.N. 193, HL, it was held that no instructions were needed in the case of a gang moving a heavy regulator, but in a similar case, *Rees v. Cambrian Wagon Works* [1946] W.N.139, CA, it was held that instructions were required for a gang moving a cog-wheel.
36   *Baker v. T.E. Hopkins & Son* [1959] 1 W.L.R. 966; *sub nom. Ward v. T.E.Hopkins & Son* [1959] 3 All E.R. 225, CA.
37   *Lewis v. High Duty Alloys* [1957] 1 W.L.R. 632.

occurrence or otherwise of previous accidents is relevant to whether or not an employer should reasonably have foreseen a similar one that has occurred. However, an employer should not wait for an accident to occur before doing something about an unsafe system of work!

Relevance of publications and statutory provisions

In certain industries, some dangers are well known and may be reported in trade journals or bulletins. In such cases, knowledge of the danger will be attributed to the employer.[38] The lessons associated with such matters should therefore form the basis of advice and instructions to employees, and prudent employers would be well advised to keep abreast of developments in relevant trade publications.

Where the circumstances of an accident are *outside the scope of the statutory provisions,* then those provisions will not be relevant to the issue of liability at common law.[39] However, the existence or otherwise of a safety provision in a code of practice for a particular type of trade may well be relevant to the *standard of care* needed at common law.[40] Notwithstanding this, an employer who has fully complied with his statutory obligations may be entitled to claim that he has also fulfilled the standard of care required of him under common law.[41]

Ordinary risks of service

Many types of work require the exercise of operations carrying an inherent and unavoidable risk. This may be called an ordinary risk of the service against which the employer cannot reasonably be expected to protect his employees.[42] Some examples of risks

---

[38] *Graham v. Co-operative Wholesale Society* [1957] 1 W.L.R. 511. *Cf. Bryce v. Swan Hunter Group* [1988] 1 All E.R. 659 (employers' knowledge of risks of asbestos).

[39] *Chipchase v. British Titan Products* [1956] 1 Q.B. 545, CA.

[40] *Franklin v. Gramophone Company* [1948] 1 K.B. 542; *N.C.B. v. England* [1954] A.C. 403, HL; *Matuszczyk v. National Coal Board,* 1953 S.C. 8; *Hartley v. Mayoh* [1954] 1 Q.B. 383, CA.

[41] *Caulfield v. Pickup* [1941] 2 All E.R. 510; *Franklin v. Gramophone Co.,* above.

[42] *Turner v. Lancashire County Council* [1980] C.L.Y. 1886 (a case involving liability to a trainee fireman); *McCafferty v. Metropolitan Police District Receiver* [1977] I.C.R. 799 (a case involving hearing damage to an officer working in a firing range).

which have been held to be ordinary risks of the service for which the employer has not been held liable are:

- An oily patch on a ship's deck.

- Dried peas on the floor of a shed.

- A protruding nail catching clothing.

- The fall of rain or wet paint making a surface slippery.

- A labourer getting his hand caught when the load he was guiding struck a stanchion, causing the load to shift.

However, if an operation involves a *real, serious and unnecessary risk*, the employer may not be able to use the excuse that the risk is one which is ordinarily found or accepted in that trade or activity.

PROOF OF NEGLIGENCE

In any claim the onus of proof normally lies with the claimant, although an admission of liability by a defendant will avoid the need to prove the claim. However, the amount of damage will have to be proved unless that is also admitted. In some cases, the court *may* allow the defendant to retract an admission of liability.[43] If the onus lies on the claimant, he must provide particulars of the claim.[44] He does not have to show that the breach of duty was the whole or main cause of the damage, only *that it materially contributed to the damage.*[45] The standard of proof is that a claimant should, on the evidence, establish a balance of probability in his favour. If he fails to do so, or the evidence leaves the balance more or less equal either way, his claim will fail. In ordinary claims the matters to be proved are:

1. The way the accident happened.

2. That what happened constituted negligence.

---

[43] Certain tests have to be applied by the court in deciding whether to allow a retraction.

[44] e.g. *Gallon v. Swan Hunter Shipbuilders Ltd., The Times,* May 18, 1995, CA (in this case, the employee could not be asked to identify each of the noxious fumes he allegedly inhaled, nor details of the employer's alleged knowledge of danger).

[45] *Bonnington Castings v. Wardlaw* [1956] A.C. 613, HL.

3.   Injury or damage was caused as a result of the negligence.

Generally, direct evidence of how the accident happened will be
needed, although it may be possible to infer this based on the
evidence as a whole. If it is proven that an accident occurred
because of a defect that would not ordinarily be present, the
claimant will not usually be required to show how the defect arose,
and the defendant will have to try and show that it arose in
circumstances that did not amount to negligence. The claimant
must prove that the defendant was negligent and the negligence
caused his injuries.[46] The injury may, in some cases be the result
of more than one cause and it may be necessary to determine the
relative contribution of different causes or of different employers
in arriving at the distribution of damages. If the claimant has also
contributed to his own injury, the damages awarded will be reduced
by a percentage relevant to the degree of his own negligence.[47]

SYSTEMS OF WORK

An important element of an employer's duty of care to his employees
is to take care to lay down a safe system of work.[48] The duty
extends to providing a safe system of work that *so far as reasonably
practicable* prevents the employee from suffering psychiatric
injury due to stress or overwork. Claims of this kind have increased
in recent years and this duty is particularly relevant if the employee
has already suffered psychiatric injury at work and conditions are
not subsequently remedied, resulting in further injury.[49]

---

[46]   *ibid.* and *Nicholson v. Atlas Steel Foundry and Engineering Co. Ltd.* [1957] 1
W.L.R. 613 (dust materially contributed to pneumoconiosis); *Yates v. Rockwell
Graphic Systems Ltd.* [1988] I.C.R. 8 (contaminated coolant caused dermatitis).

[47]   Law Reform (Contributory Negligence) Act 1945.

[48]   Consequently, an employee cannot be ordered in his action for damages to give
particulars of what he considers to be a safe system of work: *Hornett v. Associated
Octel* [1987] C.L.Y. 303.

[49]   *Walker v. Northumberland County Council, The Times,* November 24, 1994
(liability for manager's nervous breakdown). *Cf. Johnstone v. Bloomsbury Health
Authority* [1992] Q.B. 333, CA (possible liability for requiring doctor to work
excessive overtime) *Panting v. Whitbread plc* [1999] 209, Cty Ct (extent of
employer's duty). The subject of employer's liability for mental or physical injury
caused by stress at work was reviewed in *Hatton v. Sutherland, The Times,*
February 12, 2002, CA (employer liable only if he could foresee that the particular
employee could suffer injury by stress and failed to take reasonable steps to
prevent it. Proof of occupational stress was not, of itself, enough).

The question of *when* an employer is reasonably required to lay down a system of work has been the subject of debate in several court cases which appear to give rise to certain principles:

1. A system of work may not easily be applied to a case where a single act of a particular kind is to be performed but, where the operation is complicated or unusual, then a safe system of work should be provided.[50]

2. The different employers of several employees working on the same premises may all be liable, even though they are not the occupiers nor in control of the premises, if there was no system of work at all. Somebody has to co-ordinate or supervise a place to ensure that it is safe to work in.[51]

3. If an employee is sent to work on premises *not under the control of the employer,* the employer must exercise all the more care regarding his system of working.[52] A case of this kind which will be familiar to local authority enforcement staff is that of window cleaning or climbing ladders. If an employer instructs his employees in a system of work that is dangerous, he is failing in his duty.[53] There are a number of cases setting out how far an employer is under a duty to lay down a system of work for operations involving an obvious danger, many of which involve window cleaning. In one particular case,[54] a window cleaner with life-long experience lost an action for negligence because it was held that, in this particular case of such an experienced workman, the defendants had fulfilled their duty to take reasonable care for his safety and the danger which resulted in his injury was so obvious that the repetition of warnings would be likely to do more harm than good.

---

[50] *Winter v. Cardiff R.D.C.* [1950] W.N. 193, HL. *Cf. Hayes v. N.E. British Road Services* [1977] C.L.Y. 2038 (employer's duty to promote safe system of unloading lorry) and *Philip Michael Chalk v. Devises Reclamation Co. Ltd., The Times,* April 2, 1999, CA (no liability for employee's "one-off" act in lifting heavy object).

[51] *Donovan v. Cammell Laird* [1949] 2 All E.R. 82, *per* Devlin J.

[52] *Wilson v. Tyneside Window Cleaning Co.* [1958] 2 Q.B. 110, CA, *per* Parker L.J., approved by Lord Denning in *Smith v. Austin Lifts* [1959] 1 W.L.R. 100, HL.

[53] *Prince v. Carrier Engineering Co.* [1955] 1 Lloyd's Rep. 401.

[54] *Wilson v. Tyneside Window Cleaning Co.* [1958] 2 Q.B. 110, CA.

4. When a person is required to lift articles of different sizes and weights, he should not be left to decide when he needs assistance.[55] This will depend very much on the circumstances of each case.

5. Failure to ensure the co-ordination of the activities of several workmen engaged in a lifting operation constitutes an unsafe system of work.[56]

Where the standard of care relating to systems of work is in question, the standard will be that of the ordinarily prudent employer. It is for him to provide a safe system of work, although it may be necessary for the claimant to come up with a practicable alternative to that which was in operation at the time of the accident.[57]

PLACE OF WORK

An important element of an employer's duty to take reasonable care for the safety of his employees is his duty to take reasonable care to provide a safe place of work, and this includes safe means of access.[58] What is a reasonably safe place of work depends on the circumstances at the time, and there are industries where everyday conditions may be unavoidable to some extent, e.g. water on a tiled laundry floor. The fact that a hazard may be a common one will not excuse an employer from taking action to reduce the risk, particularly if the risk is serious, e.g. where the danger of slipping is so great that safeguards should be provided.[59]

The duty regarding a safe place of work includes means of access and egress, and it may well include a duty to supervise the way in

---

[55] *Per* Widgery L.J. in *Peat v. N.J. Muschamp & Co. Ltd.* (1970) 7 K.I.R. 469 at 476–477, CA. *Cf. Black v. Carricks (Caterers) Ltd.* [1980] I.R.L.R. 448, CA (not negligent to leave shop manageress to ask for customer's help, when assistant away ill).

[56] *Upson v. Temple Engineering (Southend)* [1975] K.I.L.R. 171.

[57] See *Colfar v. Coggins & Griffith (Liverpool) Ltd.* [1945] A.C. 197, HL; *Dixon v. Cementation Ltd.* [1960] 3 All E.R. 417, CA; *General Cleaning Contractors Ltd. v. Christmas* [1953] A.C. 180, HL; *Gilfillan v. N.C.B.,* 1972 S.L.T. (Sh. Ct.) 39; and *McErlean v. J. & B. Scotland* [1997] C.L.Y. 6075.

[58] *Hurley v. Sanders* [1955] 1 W.L.R. 470.

[59] *ibid.* See also *Vernon v. British Transport Commission* (1963) 107 S.J. 113 and *Newland v. Rye-Arc Ltd.* [1971] 2 Lloyd's Rep. 64.

which people enter or leave the place of work. So if an employee is injured by the uncontrolled surge of people trying to leave a room or other part of the premises, the employer may be liable.[60]

Frequent subjects of litigation are slips and trips. Not all decisions of the courts are consistent in what appear to be similar circumstances, but employers have been held liable where there was a slippery duckboard at the site of a water tap,[61] a patch of slippery ice in a cold store,[62] and where an employee had spilt tea on a floor making it slippery.[63]

In connection with work at height, it has been held that it is unsafe if a scaffold has no guard rail,[64] if the only access to a steel framework is via girders,[65] if staging is liable to swing,[66] and working on an asbestos roof without crawling boards.[67] Failing to ensure that workplace air is free from tobacco smoke to avoid the dangers of passive smoking[68] may also render the employer liable.

In a case where an employee was carrying out contract maintenance work on other premises and incurred injury as a result of defects reported several times to the occupier of those premises by his own employers, both the employer and the occupier were found to be liable. However, the degree of liability of the occupier was found to be greater because of his repeated failure to respond to the

---

[60]   *Lee v. John Dickinson*, 110 L.J. 317; *Bell v. Blackwood Morton and Sons*, 1960 S.C. 11; *Lazarus v. Firestone Tyre & Rubber Co.* [1963] C.L.Y. 2372.
[61]   *Davidson v. Handley Page* [1945] All E.R. 235, CA.
[62]   *McDonald v. B.T.C.* [1955] 1 W.L.R. 1323.
[63]   *Bell v. Department of Health and Social Security, The Times*, June 13, 1989. *Cf. Meikleham v. Greater Glasgow Health Board, The Scotsman*, June 26, 1991 (food on hospital floor; duty of employer).
[64]   *Pratt v. Richards* [1951] 2 K.B. 208.
[65]   *Sheppey v. Matthew T. Shaw & Co.* [1952] W.N. 249.
[66]   *Taylor v. Ellerman Wilson* [1952] 1 Lloyd's Rep. 144.
[67]   *Jenner v. Allen West & Co.* [1959] 1 W.L.R. 554, CA.
[68]   *Cf. Rae v. Glasgow City Council*, 1998 S.L.T. 292; *The Times*, April 22, 1997, Ct. of Sess. (OH), where the claim failed for want of pleading particularity, but it was held that s.7 of the Offices, Shops and Railway Premises Act 1963 could impose liability for not removing smoke from the air. See also *Rae v. Strathclyde Joint Police Board*, 1999 S.C.L.R. 793, OH. An employee who resigns from the employment because of the employer's failure to prevent or alleviate the risk of passive smoking may be able to claim compensation for constructive dismissal: *Waltons v. Morse and Dorrington* [1997] I.R.L.R. 488, EAT (claim succeeded, EAT citing s.2(2)(e) of the Health and Safety at Work etc. Act 1974).

defect.[69] Interestingly, in a similar case, it was held that the employer had a greater responsibility for ensuring a safe place of work for his employees than the occupier of the building where the work was carried out.[70]

PLANT AND APPLIANCES

An employer has a duty to use reasonable care to provide safe plant and appliances. Where he has done so he will not be in breach of his duty to an employee whose *authority includes the duty to select the plant to be used.*[71] Where machinery is involved, the duty in most cases will be contained in various statutory provisions but, where this is not the case, a similar duty may exist at common law.

If the operation of a machine involves an obvious danger that could easily be prevented, the employer will be negligent in failing to prevent it, notwithstanding that the *precise circumstances* that caused it could not reasonably have been foreseen.[72] Where the repeated use of a machine causes a recognisable injury, e.g. "repetitive strain injury", the employer may be liable in damages.[73] Interestingly, the employer is *not necessarily negligent* if he fails to warn the employee of the possible danger of continual repetitive movements.[74] In a fairly common case of "repetitive strain injury", e.g. typing, the employee must show that the condition is caused by and not just associated with the activity.[75]

---

[69] *Smith v. Austin Lifts Ltd. and Others* [1958] 1 W.L.R. 1958, CA.

[70] *Andrews v. Initial Cleaning Services Ltd.* [1999] 33 LS Gaz R 32, CA.

[71] *Richardson v. Stephenson Clarke Ltd.* [1969] 1 W.L.R. 1695.

[72] *Harvey v. Singer Manufacturing Co.,* 1960 S.C. 155. See also *MacDonald v. Scottish Stamping and Engineering Co. Ltd.,* 1972 S.L.T. 73n. where it was held that suggestions made by the injured worker in the case for reducing the danger were reasonably practicable and met the test of commending themselves to someone of common intelligence. His employers were therefore held liable in negligence.

[73] *Mitchell v. Atco* [1995] C.L.Y. 3722; *Ladds v. Coloroll* [1995] C.L.Y. 3723; and *Gandy v. Matesson's Walls Ltd.* [1997] C.L.Y. 2624. For the extent of medical evidence, examinations, etc. in such cases, see *Baker v. Paper Sacks* [1995] C.L.Y. 4119.

[74] *Pickford v. Imperial Chemical Industries plc* [1998] 1 W.L.R. 1189; *The Times,* June 30, 1998, HL.

[75] *Alexander v. Midland Bank plc* [1999] I.R.L.R. 723, CA.

PROTECTIVE EQUIPMENT

Where an employee is suffering from some known disability making him unusually susceptible to injury, greater care will be required in the provision of protective equipment, e.g. the question often arises as to whether goggles should be worn for a particular activity. The criterion to be adopted is whether there is a *likelihood*, as opposed to an exceptional possibility, that substantial eye damage may occur at some time.[76] The question of whether a barrier cream is required for protection against materials likely to cause dermatitis is not without doubt, but it is now well established that an employer who fails to provide adequate ear protection against noise will be in breach of his common law duty of care to his employees.[77]

Generally, the test of whether a particular item of protective equipment should be provided is just one element of the duty of care. The standard of that duty is that which an ordinarily prudent employer would take, having regard to the foreseeability of the risk, its magnitude and the practicability of taking effective precautions. What constitutes reasonable care depends on the circumstances of each case.

If an activity requires an item of protective equipment, the employer must provide it.[78] What constitutes providing something may vary from case to case, but generally this should usually involve putting the equipment in a place where it is readily available and the employee should be given clear instructions as to where and how to get it. It is of course important not only to provide appropriate protective equipment, but also to make sure it is used.

NEGLIGENCE OF OTHER EMPLOYEES

An employer is liable for the negligent acts of his employees, provided they occur in the course of their employment, and which cause injury to another employee. The standard of care required is that of a reasonably prudent employer.[79]

---

[76] *Cf. Crouch v. British Rail Engineering* [1988] I.R.L.R. 404, CA.
[77] See *Noise Control – The Law and its Enforcement*, Penn, C.N., Chapter 2, Shaw & Sons Ltd.
[78] *Finch v. Telegraph Co.* [1949] W.N. 57.
[79] *Vincent v. P.L.A.* [1957] 1 Lloyd's Rep. 103, CA.

DEFENCES

Contributory negligence

Some accidents may be caused partly by the employer's negligence and partly by the contributory negligence of the claimant. Under the Law Reform (Contributory Negligence) Act 1945, the claimant's damages will be reduced by such an amount as the court deems appropriate, provided that the defendant pleads the defence of contributory negligence.[80] An insignificant degree of contributory negligence can be ignored but, in some cases where the contributory negligence is considerable, the claimant's contribution can be assessed at 100%, thereby negating an award of damages.[81]

In general, the issue of whether an inadvertent act amounts to negligence depends on the circumstances of each case.[82]

If an employee carries out a task for which he is not qualified, and which his employer has expressly forbidden, he cannot then claim damages for any injury incurred in carrying out the task.[83] Similarly, if the injury occurs as a result of a breach of a statutory duty imposed on the individual, he may be held to have consented to the injury, even though another employee is involved.[84]

The following are useful examples of matters which are *not* contributory negligence:

1. An employee who tripped on an object on a floor while concentrating on his job has been held not guilty of contributory negligence.[85]

---

[80] Accordingly, if the defendant has been debarred by order of the court from defending the action, the court cannot, on its own initiative, reduce the plaintiff's damages for contributory negligence – *Fookes v. Slaytor, The Times,* June 19, 1978, CA.

[81] *Johnson v. Croggan & Co.* [1954] 1 W.L.R. 195. *Cf. Hewson v. Grimsby Fishmeal Co.* [1986] C.L.Y. 2255; *Bacon v. Jack Tighe (Offshore) and Cape Scaffolding* [1987] C.L.Y. 2568; and *Quant v. J.I. Case (Europe) Ltd.* [1995] P.I.Q.R. 225; [1996] C.L.Y. 4425, CA.

[82] *Hicks v. B.T.C.* [1958] 1 W.L.R. 493, CA. *Cf. Ryan v. Manbre Sugars Ltd.* (1970) 114 S.J. 492, CA. where it was held that inadvertence is not contributory negligence.

[83] *Hyland v. RTZ (Barium Chemicals)* [1975] I.C.R. 54 (and a productivity agreement allowing flexibility did not alter the position).

[84] *Imperial Chemical Industries Ltd. v. Shatwell* [1965] A.C. 656, HL; *McMullen v. National Coal Board* [1982] I.C.R. 148.

[85] *Callaghan v. Fred Kidd* [1944] K.B. 560. *Cf. Burns v. British Railways Board* [1977] C.L.Y. 357, CA.

2. A similar decision was taken in a case where an employee fell down an unguarded tank top in a momentary lapse of concentration.

3. A workman who slipped on a vertical steel ladder which was not a safe access was not guilty of contributory negligence in failing to place his feet properly on the ladder.[86]

4. A window cleaner who failed to take proper precautions was not guilty of contributory negligence, as he was only working in the way his employer expected him to do.[87] Accordingly, the established practice required or allowed by the employer can be an important factor.

LIABILITY FOR THE ACTS OF OTHERS

Employers are liable for the acts of their employees acting in the course of their employment and any of their agents acting within the scope of any authority given to them. An employer will also be ordinarily liable for the acts of sub-contractors fulfilling some part of the duty of care which he owes to his employees. Where an employer tells an employee to obey the instructions of another person, and the employee sustains injury as a result of the third person's negligence, the employer may nevertheless be made liable for the acts of the third party but may be able to claim a contribution towards the damages, having regard to the extent of the third party's responsibility for the damage. In a case involving the use of ladders, an employer agreed that one of his employees could work temporarily under the direction of another employer. The second employer failed to ensure the ladder was properly secured and the employee fell and was injured. Although the first employer was liable for his employee's safety even when he was not exercising control over him, he could not anticipate the failings of the second employer. The accident was held to be wholly attributable to the second employer.[88] It is debatable whether the

---

[86] *Graham v. Scott's Shipbuilding and Engineering Co. Ltd.*, 1963 S.L.T. 78.
[87] *General Cleaning Contractors Ltd. v. Christmas* [1953] A.C. 180, HL.
[88] *Nelhams v. Sandells Maintenance Ltd., The Times*, June 15, 1995, CA. Dictum of *Morris v. Breaveglen* [1993] I.C.R. 766, CA applied.

use of this defence would wholly exonerate the first employer in a case under the Health and Safety at Work etc. Act 1974 as the employer would be expected to check that the systems of work used by the second employer complied with his statutory responsibilities.

An employer can also be made liable to an employee injured because of defective equipment without negligence being established against him, provided that the fault can be attributed to a third party.[89] The third party will usually be the manufacturer or supplier of the equipment. The manufacturer of materials used in the employer's business may also be liable to employees for negligence.

DAMAGES

An award of damages is to compensate the plaintiff for the loss he has suffered and is measured by comparing the position of the plaintiff before and after the injury. In quantifying damages, the High Court can order the disclosure of relevant documents such as medical records. Damages will usually be assessed by a judge alone and are awarded only in respect of actual injury to health or personal property.

Under the Limitation Act 1980, an action for damages or breach of duty in respect of personal injuries must be brought within three years of either the date when the cause of action occurred or such later date as the plaintiff had "knowledge" of certain facts.[90] Accordingly, if someone takes medical advice, but is not made aware that he is suffering harm from an employer's negligence until much later, he can sue within three years of the date that he actually acquires the "knowledge". The start of the limitation period may be postponed in a case where the action is based on the fraud of the defendant, or where a relevant fact associated with the plaintiff's right of action has been deliberately concealed by the

---

[89] *Wright v. Dunlop Rubber Co. Ltd.* (1973) 13 K.I.R. 255, also reported *sub nom. Cassidy v. I.C.I., The Times,* November 2, 1972, CA.

[90] Limitation Act 1980, s.11. This defence must be taken up by the defendant, the court will not do it for him. When the defence is raised, it is up to the plaintiff to show that his first knowledge was within the limitation period.

defendant.[91] The limitation period does not then run until the plaintiff has discovered the fraud or concealment.

LIABILITY OF AN EMPLOYER TO VISITORS

The liability of an employer as the occupier of premises to visitors under common law was replaced by the Occupiers' Liability Act 1957. This lays down a "common duty of care" which is owed by an occupier[92] to all lawful visitors, but does not affect an employer's liability under health and safety legislation such as the Health and Safety at Work etc. Act 1974. The occupier need not be the owner of the premises but must have a degree of control over all or part of them, e.g. landlords are the occupiers of the common parts of buildings such as halls, lifts, landings and stairways. Failure to take steps to show that entry is not permitted may give rise to an implied permission. It might seem that enforcement officers would be protected by this duty, although the Act does say that "an occupier may expect that a person, in the exercise of his calling, will appreciate and guard against any special risks ordinarily incident to it, so far as the occupier leaves him free to do so."[93] Presumably, if this Act does not protect the interests of an enforcement officer injured on premises as a result of a failing or defect, the Health and Safety at Work etc. Act 1974 will provide a means of pursuing statutory action.

This general duty of care does not extend to any dangers caused by the actions of contractors carrying out contract work,[94] but the occupier must show that he carefully selected and monitored the contractor, although he is not expected to supervise the contractor. However, if he chooses to do so he may be liable if he then fails to do so and something goes wrong. For liability to occur under the Act, the premises must be found to be unsafe, as the Act only applies to the condition of the premises.

The Occupiers' Liability Act 1984 places a duty of care on the occupiers of premises to people other than visitors, e.g. trespassers.

---

[91] Limitation Act 1980, s.3. This includes a deliberate breach of duty owed by the defendant.
[92] Defined in *Wheat v. E. Lacon and Company Ltd.* [1966] 1 All ER 582.
[93] s.2(3)(b).
[94] s.2(4)(b).

The occupier owes the duty in respect of any risk of suffering injury on the premises by reason of any danger due to the state of the premises: if he is aware of the danger or has reasonable grounds to believe that it exists; has reasonable grounds to believe that someone may be in or come into the vicinity of the danger; and the risk is one which in the circumstances he may be expected to offer protection against.[95] He is not liable for any risks willingly accepted by someone.[96] As in the case of the 1957 Act, it is the state of the premises rather than any activities carried on in them that may give rise to liability. The Health and Safety at Work etc. Act 1974 of course covers the activities as well as the state of the premises.

---

[95] s.1(3).
[96] s.1(6).

Chapter 2

THE REGULATORY ORGANISATIONS

The regulation of health and safety is the responsibility of several agencies. This often results in an overlap of activities and the frequent need to work closely together. Accordingly, it is useful to understand their individual responsibilities and to see how the inter-relationship works. This chapter looks at those agencies, describes their responsibilities and shows how they may work together.

HEALTH AND SAFETY COMMISSION

The Health and Safety at Work etc. Act 1974 established the current health and safety organisation, a two-tier organisation headed by the Health and Safety Commission (HSC). The Commission is empowered to give directions[1] to its operational arm, the Health and Safety Executive (HSE). The Commission, a body of up to ten people, was appointed by the then Secretary of State for Transport, Local Government and the Regions and in turn appointed the Executive.[2]

The Commission has wide statutory duties, e.g. to ensure that the general purposes of the Act are fulfilled, to arrange for research, and to give advice and information.[3] Its functions are subject to directions from the Secretary of State for Employment.[4] Its responsibilities include proposing health and safety legislation and standards to Ministers, relying on advice from the Executive and the results of scientific research carried out in the Executive's own laboratories or contracted out research. It has powers to institute agency agreements for government departments or others, and to carry out functions on its own behalf or on behalf of the Health and Safety Executive,[5]

---

[1] Health and Safety at Work etc. Act 1974, s.11(4).
[2] The D.T.L.R. no longer exists. The relevant department is the Department for Work and Pensions.
[3] Health and Safety at Work etc. Act 1974, s.11.
[4] *ibid.,* s.12. Now Secretary of State for Education and Employment.
[5] *ibid.,* s.13. It can also carry out work by agreement on behalf of government departments or public authorities.

to take and pay for expert advice,[6] to direct the holding of investigations or inquiries into accidents or other occurrences,[7] and to approve and issue codes of practice. Failure to comply with an approved code of practice is admissible evidence in criminal proceedings for breach of statutory duties.[8] The Commission also consults widely with organisations representing professional interests in health and safety, trades unions, scientific and technical experts, and business representatives. It does this through a series of Advisory Committees, Boards and Councils on which these various groups are represented. It has a specific duty to maintain the Employment Medical Advisory Service (EMAS), which provides advice on occupational health matters. It does this in practice through the Executive.

A primary task of the Commission is to progress the results of a strategic appraisal of the health and safety framework in Britain. *Revitalising Health and Safety*[9] is a ten-year strategy statement published by the then DETR and the Health and Safety Commission following an extensive consultation exercise in 1999. It takes a fresh look at the health and safety policy and framework and seeks to set a new agenda for the first quarter of the new Millennium. It contains a 10-point strategy supported by a 44-point action plan designed to provide incentives and practical support to employers, together with a range of measures intended to address the problem of employers who fail to meet their health and safety responsibilities. The government clearly thinks that the basic framework of the Health and Safety at Work etc. Act 1974 has stood the test of time and seeks to build on that by "adding value" to the current system. For example, while accepting the need for appropriate enforcement and deterrence, it wishes to promote voluntary compliance and models of excellence, building on the successful partnerships that have taken place over many years between employers, employees, trades unions and consumers. Examples of such good practices are included in various parts of this book.

The strategy statement sets out how the government and the

---

[6]   Health and Safety at Work etc. Act 1974, s.13(1)(d).
[7]   *ibid.,* s.14.
[8]   *ibid.,* ss.16-17(2).
[9]   *Revitalising Health and Safety*, Strategy Statement, June 2000, DETR.

Commission will work together and publishes the first overall targets for the health and safety system:

- To reduce the number of working days lost per 100,000 workers from work-related injury and ill-health by 30% by 2010.

- To reduce the incidence rate of fatal and major injury accidents by 10% by 2010.

- To reduce the incidence rate of cases of work-related ill health by 20% by 2010.

- To achieve half the improvement under each target by 2004.

These targets are underpinned by a 10-point strategy which is intended to set the direction for the health and safety system up to 2010. It seeks to emphasise the importance of "promoting better working environments to deliver a more competitive economy, motivating employers to improve their health and safety performance, and simplifying over-complicated regulations." The strategy statement involves:

1. Promoting better working environments – involving motivated workers, competent managers and the adoption of *best practice* rather than minimum standards.

2. Promoting the contribution of employees through ensuring that they understand their own responsibilities and the benefits associated with a strong health and safety culture.

3. Maintaining occupational health as a top priority by ensuring better compliance with the law, securing continuous improvement and the availability of the right knowledge and skills.

4. Getting the positive involvement of small firms by showing how they can benefit from effective health and safety management, simplifying the law where possible and encouraging firms to seek advice.

5. Reforming the compensation, benefits and insurance system

to encourage employers to improve their health and safety performance.

6. Encouraging more self-regulation, especially in the 3.7 million businesses with less than 250 employees.

7. Encouraging greater worker involvement in health and safety management through partnerships on health and safety issues.

8. Getting public bodies to lead by example by demonstrating their use of best practice in health and safety management.

9. Providing education on health and safety skills and risk management from primary school to the workplace.

10. Designing health and safety properly into processes and products.

Action plan

Revitalising Health and Safety contains a 44-point action plan intended to give effect to the strategy. These points are detailed and can readily be referred to in the *Revitalising* document.[10] There are a number of them which it might be useful for local authorities to consider in addressing their own strategies for the regulation of health and safety. These can be summarised as follows:

1. *The HSC will review the health and safety reporting regulations.* Whilst most fatalities are reported to the Health and Safety Executive or local authorities, only 47% of reportable incidents are actually reported. Accordingly, a key element of local authority work may need to involve investigation of why there is so much under-reporting of accidents or near misses that should be reported, with enforcement of the regulations and targeted education forming an element of their local strategy.

2. *Health and safety management standards.* The HSC will advise Ministers what steps can be taken to enable companies to check their health and safety management arrangements

---

[10]   pp.20-41.

against an established yardstick. Some local authorities already provide written advice and guidance on what they expect of those businesses subject to local enforcement. This often includes advice on self-regulation.[11]

3.   *"Naming and shaming" companies and individuals convicted of offences.* The HSE will draw to public attention trends in prosecutions, convictions and offences, including naming convicted companies and individuals. Local authorities frequently complain about low penalties imposed by the courts for health and safety convictions. Accordingly, a similar approach to that of the HSE may well be a useful local or regional initiative, coupled with appropriate training and education programmes, that could lead to greater awareness and understanding by employers of their responsibilities.

4.   *Directors' responsibilities.* The HSE has published advice on directors' responsibilities for health and safety.[12] Those basic requirements should of course be included as part of a local authority training and education programme. Such information could also usefully form part of any basic information given to new businesses as an introduction to their legal responsibilities towards their employees.

5.   *Health and safety checklists.* The *Revitalising* document outlines a health and safety checklist to be the subject of consultation.[13] That checklist might also be issued as an aide memoire to the employers of local authority enforced premises. It might usefully read along the following lines:

   •   Have you developed a health and safety plan and identified your highest risk activities?

   •   Do your safety policy and risk assessments comply with the law? Who is named as responsible for health and safety?

---

[11]   e.g. *Guidance on undertaking COSHH assessments,* City of Bradford Metropolitan District Council, December 2002.

[12]   See "Directors' responsibilities for health and safety", INDG343, July 2001, HSE, and also a research report "Health and safety responsibilities of company directors and management board members", 2003, HSE.

[13]   *Revitalising Health and Safety,* Strategy Statement, Annex B, June 2000, DETR.

- How do you ensure people know of your commitment to health and safety?

- What steps do you take to safeguard members of the public who visit your premises?

- How do you monitor, review and audit your health and safety performance?

- How do you consult, advise and motivate staff on health and safety?

- Do your risk assessments and control methods reflect individual abilities and needs?

- How many RIDDOR reportable injuries/diseases/ dangerous occurrences have you had and reported in the last 12 months?

- Have you got proper written records of your risk assessments, training, accidents, dangerous occurrences and near misses?

- Are all incidents investigated so that lessons are learnt and relevant risk assessments reviewed?

6. *Worker contact on health and safety issues.* The HSE is taking action to publicise the right of employees to contact them on health and safety matters, particularly cases of unsatisfactory practices.[14] Local authorities could usefully offer all employees, both internal and external, an *advice line* coupled with internal procedures to protect their anonymity.

7. *Effective guidance for small firms.* Although the HSE provides many useful guidance notes, 75% of small firms suggested that appropriate advice for their specific needs was not available. As a result, the HSE is reviewing its guidance. However, many of these small firms are also local authority controlled and there may be a case for local authorities joining

---

[14] Protection of such "whistleblowers" is afforded by the Public Interest Disclosure Act 1998.

forces to produce relevant guidance for certain types of
activity, possibly based in some cases on information already
produced by individual authorities.[15]

8. *The HSC will work with local authorities to propose an
   indicator against which the performance of local authority
   enforcement and promotional activity can be measured.*
   Ninety-seven percent of the consultation responses to the
   question about whether more could be done to raise the profile
   of local government health and safety enforcement activity
   had the answer "yes". Dissatisfaction with the performance of
   local government is being reflected in a review of that
   performance by the HSE. The Chairman of the Commission
   and the Head of the HSE's Local Authority Unit have already
   warned local authorities that they need to give a much higher
   priority to health and safety performance if they wish to avoid
   their duties being transferred to the Executive.[16] This attack
   on a traditional role of local authorities reflects to some degree
   the political priorities of some councils, and a lack of real
   interest of others in this essential work. Either way, it is up to
   health and safety professionals to take action by reviewing
   their approach to this area of work, improving efficiency and
   effectiveness and persuading elected members to give health
   and safety the right priority.

9. *The development of proposals for sharing with health and
   safety regulators information about business start-ups.* There
   has been no uniform method of referring new business start-
   ups, but there is a great deal of information about new business
   activities that can be obtained from planning approvals,
   commercial rates information and general area surveys that
   can inform local authorities of new businesses that require
   health and safety inspections.

10. *The HSE and the government will act in partnership to
    increase the number of staff secondments arranged between
    the Executive and central or local government, industry or*

---

[15] e.g. *Call Centres,* City of Bradford Metropolitan District Council, August 2002.
[16] See "Health and Safety Matters", *Environmental Health Journal,* p.238, August
      2003.

trades unions. The current appointment of a former local government chief officer as Head of the HSE's Local Authority Unit is a useful step in improving liaison and partnership arrangements.

The strategic plan

The Commission has produced a plan setting out how it proposes to deliver the action points for which it is responsible. Its *Strategic Plan 2001-2004*[17] contains its strategic direction for health and safety in Great Britain over that period. There are four broad considerations to the Commission's thinking:

1. Its plan concentrates on what it aims to achieve, not how it plans to do it.

2. It seeks to make all areas of work safer and healthier.

3. It will continually examine its own effectiveness.

4. It wishes to work with everyone with an interest in health and safety, including employers, employees, trades unions, trade associations, small firms, local authorities, consumer bodies and others.

The strategic plan also makes reference to a long-term strategy to improve occupational health, *Securing Health Together.* This is a central element of the *Revitalising Health and Safety* strategy statement, committing the HSC, HSE, the government and local authorities, to work together to achieve the health-related targets in *Revitalising Health and Safety,* and the following additional targets by 2010:

- A 20% reduction in ill health to members of the public caused by work activity.

- Everyone currently in employment but off work due to ill health or disability to be, where necessary and appropriate, made aware of opportunities for rehabilitation back into work as soon as possible.

---

[17] *Health and Safety Commission – Strategic Plan 2001-2004*, Health and Safety Executive. A summary of the plan is also available.

- Everyone currently not in employment due to ill health or disability to be, where necessary and appropriate, made aware of and offered opportunities to prepare for and find work.

The strategic plan sets out a framework for its activities; a number of priority work programmes; the work it will undertake to deal with the risks in major hazard industries; the extent of its proposed enforcement action; its policy, technical and information activities; and the financial and manpower requirements to achieve the planned outputs. By 2003/2004 the Commission anticipated a budget of £212,900,000 and a combined HSC/HSE staffing level of 4,237.

The commission has further developed its strategy. This is set out in "A strategy for workplace health and safety in Great Britain to 2010 and beyond"[18] and takes forward the *Revitalising Health and Safety* strategy statement of June 2000. The key challenge is "how to make appropriate risk management relevant to the modern and changing world of work". The extended strategy reflects a number of issues, including:

1. The boundaries and direction of the health and safety system need to be set.

2. HSE and local authority resources are limited and need to be targeted to the best effect.

3. The traditional interventions may be less effective when dealing with health than when dealing with safety.

4. Many organisations are fearful of confronting the HSE or local authorities.

The strategy contains the HSC and HSE's continuing aims.[19] There are also some new aims:

- To develop new ways to establish and maintain an effective health and safety culture in a changing economy, so that all employers take their responsibilities seriously, the workforce is fully involved and risks are properly managed.

---

[18] MISC 644, February 2004, HSE.
[19] As contained in the *Revitalising* strategy.

- To do more to address the new and emerging work-related health issues.

- Achieve higher levels of recognition and respect for health and safety.

- Exemplify public sector best practice in managing resources.

There is an emphasis on the HSE and local authorities working more closely together; centrally co-ordinated programmes of activity for certain activities and large organisations; more sensible and better understood divisions of enforcement; sharing of training, intelligence and expertise; and a review of the effectiveness and value of the existing HSE and local authority liaison arrangements.

Amongst other things, there will be a simplification of the concept of risk assessment; the promotion of greater employee involvement in health and safety management; improvement of communication in small businesses to try and overcome the fear of enforcement action; clarifying priorities and better targeting of the available resources into areas of greater risk; a commitment to meeting the *Revitalising* targets; and development by the HSE of a business improvement programme to ensure a culture of continuous improvement.

HEALTH AND SAFETY EXECUTIVE

The Health and Safety Executive is a body of three people appointed by the Commission with the approval of the Secretary of State for Transport, Local Government and the Regions. It advises and assists the Commission and has its own specific duties, primarily the enforcement of health and safety law. The Health and Safety Executive staff includes inspectors, scientific and technical staff, and policy advisors.

The Health and Safety (Enforcing Authority) Regulations 1998[20] allocate the type of premises subject to control by the HSE. The HSE is responsible for enforcement in over 740,000 premises, mainly those involving higher than average risk to employees. The HSE's offices are organised on a regional basis across Britain.

---

[20] S.I. 1998 No. 494.

The HSE provides the Commission with advice on policy and legislation through its strategy and support divisions. That process involves consultation on the Commission's behalf with its own expert staff, a variety of advisors in industry and commerce, a network of advisory committees which recommend standards and guidance, and consultation with local authorities through the HSE/ Local Authority Enforcement Liaison Committee (HELA).

Any proposals for new legislation or codes of practice are preceded by consultation documents issued by the HSC. These are publicly available to anyone with a relevant interest and are widely circulated to those organisations which may be affected by the proposals. The consultation procedure is also used to consider proposals originating from the European Union (EU). Where EU proposals are concerned, the HSC seeks ultimately to maintain or improve established UK standards.

The Commission's consultation and liaison arrangements involve engineering institutions, universities, and scientific and professional societies, e.g. Chartered Institute of Environmental Health. The HSC and HSE also work closely with international organisations such as the European Union, the European Agency for Health and Safety at Work, the World Health Organisation and the Organisation for Economic Co-operation and Development.

The Commission's activities are contained in its strategic plans[21] which set out its medium-term proposals and its annual business plans.[22] The business plans reflect, amongst other things, the targets of the strategic plans. Local authorities are expected to prepare their own plans to take account of the views of the Commission that are contained in its plans.

HEALTH AND SAFETY EXECUTIVE/LOCAL AUTHORITY ENFORCEMENT LIAISON COMMITTEE (HELA)

The main HELA committee meets three times a year under the joint chairmanship of the Deputy Director General of the Health and Safety Executive (HSE) and a senior local authority manager

---

[21]   The latest being *Strategic Plan: 2001-2004*, HSC.
[22]   The latest being *Business Plan: 2003-2004*, HSC.

(LA) nominated by the Local Authority Associations (LAAs). They host and chair meetings alternately. Its membership consists of senior staff from health and safety, trading standards and fire departments representing the LAAs and nominated by the Local Government Association (LGA) in England and Wales and the Convention of Scottish Local Authorities (COSLA). It also includes appropriate HSE officers.

The main committee deals with major policy and strategic issues relating to local authority health and safety performance. It also monitors the HELA strategy which sets out local authority input to the HSC's aims for improved health and safety performance. It also reports on its achievements and gives details of local authority enforcement in its annual report and its National Picture report. It produces an Audit Protocol for the inter-authority auditing of local authority enforcement management, and issues guidance under section 18 of the Health and Safety at Work etc. Act 1974. This contains the broad details of how the HSC expects local authorities to enforce health and safety legislation.

LOCAL AUTHORITY UNIT (LAU)

The LAU is part of the Policy Divisions of the Health and Safety Executive and reports to the Deputy Director General (Operations). It has been subsumed into a Policy and Operations Division. The head of the LAU reports to a senior HSE official who in turn reports to the Deputy Director General (Operations). Traditionally, it has enjoyed a special relationship with the Commission as well as supporting HELA by providing a secretarial role. Although, at the time of writing, the whole of the current system is under review, the LAU currently co-ordinates the liaison arrangements between the HSE and local authorities to meet the Commission's objectives. It works with the HELA committee to produce national advice, information and guidance to local authorities, develops enforcement policy and provides local authorities with training and support. In particular, it seeks to:[23]

• Support the work of HELA.

---

[23] See HELA LAC 48/2 (rev 2), October 2000.

- Develop and promote the partnership between the Health and Safety Commission/Executive and maintain an effective channel of communication between them.

- Promote the principles of consistency, transparency, targeting and proportionality.

- Enhance the standing of local authority health and safety enforcement among local authorities, the Commission and Executive, government, business, trades unions and consumers.

- Work with HELA to provide national advice, information and guidance to local authorities.

- Develop complementary and consistent health and safety policies between the authorities and between them and the Executive.

- Provide intelligence and advice to all of the enforcement bodies.

- Maintain a national database of health and safety information relating to the local authority enforced sector.

- Maintain a dialogue with relevant business sectors and trades unions.

ENFORCEMENT LIAISON OFFICERS

At an operational level, one of the most useful forms of liaison is through Enforcement Liaison Officers. These are operational Principal Inspectors in the Health and Safety Executive and are located in most main Executive offices.[24] They and their teams are expected to establish links with local authority staff to:[25]

- Support local authorities in meeting the HELA strategic plan.

- Develop joint initiatives, especially where there is shared enforcement.

---

[24] See HELA LAC 48/2 (rev 2), October 2000 for main office details.
[25] *ibid.*, paras. 13-18.

- Provide operational support and advice.

- Act as a channel of communication.

- Advise on, and resolve, enforcement allocation matters.

- Carry out formal local transfers and assignments.

Enforcement Liaison Officers will also provide reactive help. They provide advice on all technical standards, enforcement matters and can make available expert medical, laboratory and other advice. There is also an out-of-hours duty system for dealing with emergencies and serious incidents.[26]

The commitment to partnership arrangements with local authorities is contained in a *Memorandum of Understanding*.[27] Amongst other things, the memorandum commits the HSE and local authorities to a shared undertaking to controlling risks to people's health and safety, maintaining effective channels of communication, and working effectively and efficiently. Local authorities are expected to demonstrate their commitment by:

(a) making adequate arrangements for the enforcement of health and safety law in line with their duties under section 18 of the Health and Safety at Work etc. Act 1974, in accordance with the Health and Safety Commission's general direction and its specific LA Guidance and Enforcement Policy Statement[28] and having regard to local needs and circumstances;[29]

(b) providing appropriate and adequate levels of training and development to ensure the continued professionalism and competence of enforcement officers;

(c) through HELA, promoting best practice in local authority enforcement.

---

[26] HELA LAC 48/2 (rev 2), October 2000, telephone numbers at para. 21.
[27] Between the Health and Safety Executive, the Local Government Association and the Convention of Scottish Local Authorities, March 2001.
[28] See Appendix to section 18 guidance.
[29] This phrase is intended to provide local authorities with flexibility, although it is not intended to result in the minimal health and safety activity of some of them.

For its part, the HSE is committed to:

(a) continue to provide a focus for local authority health and safety enforcement activity within the HSE through its Local Authority Unit (LAU) and HELA;

(b) develop appropriate guidance and advice, in consultation with local authorities, on the execution of their functions and responsibilities and give practical support through the Enforcement Liaison Officer system;

(c) where possible, develop joint training initiatives and share training and development expertise with local authorities.

Communication is expected to be open and supportive of each organisation. A standing conference on health and safety enforcement in the local authority sector has been established[30] and the HSE and local authorities develop joint initiatives and monitor the effectiveness of the demarcation of their responsibilities. The *Memorandum* is revised as appropriate following advice from HELA to the local authority associations.

Inter-authority auditing is an essential part of the process of ensuring enforcement consistency and effective liaison between local authorities. It seeks to identify good practice, consistency and continuous development.[31] HELA has produced an *Audit Protocol*[32] consisting of guidance documents and questionnaire, and intended to be used within county, regional or other liaison groups or by external auditors.

LOCAL AUTHORITIES

Local authorities are responsible for enforcing the Health and Safety at Work etc. Act 1974 in about 1.2 million premises, generally in the lower risk employment sectors. There are 410 authorities covering England, Scotland and Wales. Under section 18(4) of the Act, local authorities are required to "make adequate

---

[30]   The annual Lancaster Symposium.
[31]   See HELA LAC 23/19, January 2002 for more details.
[32]   *Auditing Framework for Local Authorities' Management of Health and Safety Enforcement*, HSE, September 2002.

arrangements" for the enforcement within their area of the relevant statutory provisions and must perform that duty and any other functions conferred on them by any of the relevant statutory provisions *in accordance with such guidance as the Health and Safety Commission may give.*[33]

These guidance notes contain the broad principles that the HSC expects local authorities to adopt in enforcing heath and safety legislation. They include a published statement of enforcement policy and practice; a system for prioritised planned inspections; a service plan setting out an LA's priorities, aims and objectives for enforcement; the resources to investigate workplace accidents and complaints; arrangements for comparing their performance with other LAs; properly trained staff; and co-ordination arrangements under the Lead Authority Partnership Scheme.[34]

The guidance is not intended to be so prescriptive that it fails to take account of local needs and circumstances, but an inter-authority protocol[35] issued by HELA enables the Commission and local authorities to review and monitor the performance of individual local authorities. The section 18 guidance is supplemented by guidance from HELA through Local Authority Circulars (LACs) and other advisory documents. All of that guidance should be considered by LAs in deciding how to comply with their obligations under section 18.

Local authorities must ensure that they provide sufficient resources for enforcement of health and safety legislation to comply with their statutory duties under section 18 of the Health and Safety at Work etc. Act 1974. The actual level of such resources that are provided will inevitably depend both on what is needed to meet the requirements of the Commission in its annual *Business Plans*[36] and the priorities of each enforcing authority. The political priorities of local authorities on health and safety matters may be different from those of the Commission. A greater emphasis on housing,

---

[33] "Health and safety in local authority enforced sectors", *Section 18 HSC Guidance to Local Authorities*, HSC, October 2002.

[34] See pp.45-49 for more details.

[35] Contained in HELA LAC 23/19, January 2002.

[36] The latest at the time of writing being *Business Plan: 2003-2004*, HSC.

education and social services issues is, after all, likely to produce more votes! As a result, there is a wide variation in the level of health and safety enforcement across the country. That variation was revealed in the HSC's 2002 report on the national picture on enforcement.[37] In the report, the HSC stated that it was "concerned over the growing trend among local authorities to reduce the priority given to health and safety enforcement resulting in a 10% drop in inspection rates". The figures revealed:

- A fall in the number of local authority inspectors for the fifth consecutive year – a fall of almost 12% since 1998/9.

- An 11% fall in inspections over 1999/2000 and a 20% decrease since 1998/9.

- A large variation between different local authorities in their levels of inspections, investigations, enforcement notices served, and the number of health and safety staff.

- Some authorities had a high ratio of visits to premises whilst others undertook almost no health and safety work at all.

- Whilst around 25% of those reporting on their activities investigated every reported injury to a worker, 17 investigated less than 10% of reported injuries. Indeed, one authority only investigated 3% of the reported injuries, over 10% of which were major injuries.

- One authority served an enforcement notice for every eight visits, whilst another served only one from its 1,116 visits.

- The ratio of inspectors to the number of premises controlled varied from around one per 275 to only one part-timer for over 3,000. The largest fall in the ratio of inspectors to premises controlled occurred in the London boroughs.

The trade union UNISON has produced an interesting and useful assessment of the HSC report[38] which provides a more detailed

---

[37] *Health and safety in local authority enforced sectors*, HELA Annual Report 2002, November 2002, HSE.

[38] *Safety Lottery – how the level of enforcement of health and safety depends on where you work*, September 2003.

analysis and also identifies some of the poor performing authorities. There is clearly a wide variation in the levels of inspection, investigation and use of enforcement powers that needs to be addressed. Thirty local authorities failed to submit an annual return and, oddly, there is no legal obligation to do so. The lack of consistency in enforcing the Health and Safety at Work etc. Act 1974 is a major problem, particularly at a time when the number of local authority controlled premises is increasing, especially in high risk areas such as care homes for the elderly. The Head of the HSE's Local Authority Unit is reported as saying "This is a worrying trend ... Local government must recognise its responsibility for health and safety enforcement and make adequate arrangements to deliver this duty", adding that "there are many who seem unconcerned about reducing accidents and ill health in the workplace".

Although the HSC does not, and realistically cannot, specify the performance required of individual local authorities in its *Business Plan,* the poor performance of some authorities in meeting their legal enforcement obligations under section 18 of the HSW Act could lead to the Secretary of State invoking his powers under section 45 of the Act. He may, after considering a report from the HSC, cause a local inquiry to be held[39] and, if satisfied that the local authority has failed to perform any of its enforcement functions, he may make an order declaring the authority to be in default and ordering them to perform their enforcement functions in a specified manner and in a specified period of time.[40] If the authority fails to comply with an order under section 45, the Secretary of State may enforce it or make an order transferring the enforcement functions of the authority to the Health and Safety Executive.[41] This is an implied threat already issued by the HSC[42] but made with reference to the possibility of all health and safety at work enforcement being transferred to the HSE.[43]

---

[39]   s.45(2).

[40]   s.45(3)(4).

[41]   s.45(5).

[42]   See comments attributed to the Chairman of the Commission on p.91, Chapter 3.

[43]   How such an arrangement would work is difficult to see at present as the HSE is itself significantly under-resourced.

Section 18 guidance to local authorities

The mandatory guidance issued by the HSC under section 18 of the Health and Safety at Work etc. Act 1974 consists of six Guidance Notes. These reflect the elements the HSC considers essential for a local authority to discharge its functions under the Act. They are set out below:

A published statement of enforcement policy and practice
Local authority enforcement policies are required to be consistent with the HSC Statement on Enforcement Policy.[44] The HSC's own statement takes account of the government's "Enforcement Concordat" which sets out the principles of good enforcement practice. Local authority policies should not only take these documents into account, but also state that they have done so. A good example of a comprehensive enforcement policy is that produced and issued by Birmingham City Council Environmental and Consumer Services Department in February 2003.[45] The department has adopted the six principles of "good enforcement" set out in the government's Enforcement Concordat, these being:

1. Performance will be measured against published standards.

2. There will be openness in dealing with businesses and others.

3. Enforcement officers will be helpful, courteous and efficient.

4. Complaint procedures will be published.

5. Enforcement decisions will be proportionate to the circumstances.

6. Enforcement officers will strive for high standards of consistency.

The policy reflects the Best Value Performance Indicator (BVPI) 166 and is updated to ensure continued compliance with that indicator. It takes account of the guidance issued to prosecutors by

---

[44] Annex 1 of the *Section 18 Guidance to Local Authorities*, HSC.
[45] A District Audit against the Best Value Indicator [BVPI 166] *"Score against a checklist of enforcement best practice for Environmental Health and Trading Standards"* gave the service a 100% compliance rating.

the Crown Prosecution Service in the *Code for Crown Prosecutors*;[46] states the factors to be considered before formal enforcement action is taken; identifies liaison arrangements with other regulatory bodies, including joint enforcement arrangements where they exist; publishes the policy on the City Council's website;[47] supports the policy with detailed operational guidance procedures; and takes full account of legislation dealing with human rights, e.g. the Data Protection Act 1998 and the Regulation of Investigatory Powers Act 2000. The detail of the policy covers general enforcement principles; keeping alleged offenders, witnesses and others informed of progress; how decisions are made on the appropriate level of enforcement action; determining whether a prosecution or formal caution is viable and appropriate; the level of decision-making on what enforcement action is taken; liaison arrangements with other regulatory bodies and enforcement agencies such as the police and Health and Safety Executive; how the interest of those affected by offences will be taken into account; and how human rights will be protected. The policy is reviewed annually.

The policy is supported by a policy guidance document which outlines how the department will pursue its Enforcement Strategy. This identifies a risk-based inspection approach; the criteria for investigating accidents which is based on a departmental Quality Procedure;[48] the use of education as an aid to improving health and safety; the use of informal action and advice; prosecution criteria; liaison arrangements in the case of a death at work; the use of publicity to draw attention to prosecutions and convictions; and the use of performance measures to maintain control of their activities.

A system for prioritised planned inspection activity according to hazard and risk
The section 18 guidance expects local authorities to operate priority planning systems based on their assessment of the risks associated with different work activities. This should enable

---

[46]   Crown Prosecution Service, 2000.
[47]   www.birmingham.gov.uk/publichealth.
[48]   HS-2 "Accident Investigation".

resources to be directed at the areas of greatest need. There are numerous examples of well considered plans on local authority websites.

A service plan detailing the LA's priorities and its aims and objectives for enforcing health and safety
There are a number of comprehensive and well thought out service plans in existence which reflect the HSC's essential elements. Chichester District Council, for example, publishes a shortened version for members of the general public.[49] It sets out what the council intends to do in respect of routine visits, special projects and initiatives, accident investigations, advice to businesses, health and safety promotion, training and accident investigation, and the investigation of complaints. It describes how it will use its resources together with the arrangements for monitoring and reviewing its performance. The more detailed plan reflects the findings of a Fundamental Service Review in 2001/02 which resulted in a five-year continuous improvement plan designed to improve service delivery. The plan contains its health and safety aims and objectives; its commitment to inter-authority auditing of the health and safety function at least once every five years; its staff training arrangements; its links to the council's corporate objectives; and its targets to meet Local Performance Indicators and the premises inspections and visits it will carry out. These are based on the inspection frequencies set out by HELA.[50] The Work Plan establishes that the health and safety activities will fully reflect HSC and HELA guidance.

The capacity to investigate workplace accidents and to respond to complaints by employees and others against allegations of health and safety failures
Service planning should be a response to both historical records of complaints and accidents, but should also reflect trends in these matters as well as changes that might be anticipated in response to additional demands, e.g. new premises allocated to local authorities for enforcement or increased demands from the HSC through

---

[49] *Health and Safety Service Plan 2002/2003*, available on its website www.chichester.gov.uk
[50] HELA LAC 67/1 (revised).

revisions to its section 18 guidance. Most good service plans will show trends and offer reasoned explanations for new proposals.

Arrangements for benchmarking performance with peer local authorities
The section 18 guidance anticipates that quality audits will form a key part of local authority activity, measuring performance against agreed standards and benchmarks, to ensure that policies are being adhered to and that the aims and objectives of the organisation are being achieved. However the audits are conducted, the results should give rise to an action plan for continuous improvement. The HSC also expects that auditing and benchmarking activities will produce consistency in approach between authorities. The HSC has produced an "Auditing framework for local authorities' management of health and safety enforcement" which is available on the HSE website. The use of this framework will help to assure authorities that they are complying with the section 18 mandatory guidance. The HSC expects all local authorities to undergo an audit of their management of health and safety performance at least once every five years. It may review such reports and any resultant action plans to see where improvements may be required. The HSC has identified a number of authorities who have carried out audits and produced results that meet its expectations. These are contained in the HELA Annual Report 2000.[51] Those authorities are willing to share their experience of the auditing process and the subsequent service reviews. However, despite the HSC's ambitions for the auditing process, it should not be forgotten that consistency in approach and service development will also depend on the political will in the face of competing priorities! The analysis of performance on behalf of UNISON of the latest report on the national picture[52] suggests there is a long way to go.

Provision of a trained and competent inspectorate
The HSC expectation is that all inspectors will have the ability to perform all the activities needed to meet required levels of performance. This includes organisation and planning of work and

---

[51] Published by the HSE.
[52] *Safety Lottery – how the level of enforcement of health and safety depends on where you work*, September 2003.

undertaking enforcement action in accordance with legal procedures, as well as developing the interpersonal skills necessary to deal with a wide range of situations which may range from routine explanation of minor health and safety issues to an employee, to investigation of a fatality at work and the difficult task of dealing with bereaved and traumatised relatives.

The HSC considers that the competence elements for health and safety regulators published by the Employers' National Training Organisation represent the core competences needed by inspectors to exercise their powers under the HSW Act. These are to:

1. Identify the plans and priorities of the regulatory authority for work-related health and safety, and contribute to them effectively.

2. Inspect duty holders, work sites and activities for the purposes of work-related health and safety regulation.

3. Investigate work-related accidents, incidents, ill-health reports and complaints for the purposes of health and safety regulation.

4. Plan and gather evidence for the purposes of work-related health and safety regulation.

5. Enforce statutory provisions and brief a prosecutor for the purposes of work-related health and safety regulation.

6. Enforce statutory provisions and present guilty pleas in a magistrates' court for the purposes of work-related health and safety regulation.

7. Draft and serve notices or other statutory enforceable documents for the purposes of work-related health and safety regulation.

8. Influence health and safety duty holders and others for the purpose of work-related health and safety regulation.

9. Improve work-related health and safety through promotional activities.

A proper training regime is an essential part of any service plan. Most competent authorities will have a formal annual staff appraisal system which should identify any individual training needs as well as general training requirements, e.g. resulting from new legislation. Theoretical training should always be followed up with supervision by a trained and competent colleague.[53] There are plenty of good training organisations available and the HSE is willing to provide guidance and participate in joint training exercises in appropriate cases.[54]

Arrangements for liaison and co-operation in respect of the
Lead Authority Partnership Scheme (LAPS)
The requirements in respect of this scheme are contained in the section 18 guidance[55] and in many ways mirror the "home authority" principle established by the Local Authorities Co-ordinating Body on Food and Trading Standards (LACOTS) in 1981. An information booklet on LAPS[56] provides guidelines for local authorities on the operation of the scheme and information for businesses on how to participate in the scheme. The Lead Authority Partnership Scheme aims to:

(a) promote consistency in the health and safety enforcement of organisations with multiple outlets within the local authority enforced sector;

(b) improve health and safety management systems within these organisations.

The basis of such schemes is that enforcing authorities enter into voluntary partnerships with one or more organisations with multiple outlets in a number of local authority areas. These could include companies, charitable organisations, trade or other associations. There are currently over 100 partnerships and many have been in operation for a number of years since the scheme was established in 1991.

---

53 For further details see Chapter 6, pp.293-305.
54 Through the nominated Enforcement Liaison Officers.
55 Guidance Note 6.
56 *Lead Authority Partnership Scheme – liaison on health and safety in the local authority enforced sector*, November 2002, HSE.

The main elements of the scheme are:[57]

1. An optional Agreement of Intent signed by the participants. The lead authority is usually, but not necessarily, the authority in whose area the organisation's head office is located and acts as a point of liaison for other authorities on health and safety matters affecting the partner organisation. The lead authority familiarises itself with the health and safety management systems of the partner organisation through a joint evaluation exercise. This should then lead to the production of an Action Plan setting out requirements to ensure full compliance with the organisation's statutory obligations. This information is shared with the participating authorities. If the partner organisation fails to fulfil the requirements of the Action Plan, it may be subject to legal action. *Partnership is not an excuse for non-compliance!*

2. All health and safety enforcing authorities should normally contact, and if necessary liaise with, the relevant lead authority before taking formal enforcement action, except in the case of immediate danger: as soon as practicable after serving a prohibition notice; when significant shortcomings are identified in a participating organisation's agreed policies and procedures which the enforcing authority feels should be challenged at national level; or following any on-site investigation into an occurrence reportable under the Reporting of Injuries, Diseases and Dangerous Occurrences Regulations 1995.

3. Nationally, local authorities are expected to benefit from access to information, advice and support from participating organisations, improved use of their resources and more consistency of enforcement.

4. For lead authorities, advantages are seen in dealing with major employers; the ability to highlight some issues nationally; improving the health and safety management and influencing

---

[57] See *Lead Authority Partnership Scheme – liaison on health and safety in the local authority enforced sector* for full details. See also HELA LAC 44/3, September 2000 for further information including the answers to some common questions.

the decision making of the partner; and raising the profile of the authority.

5. The partner organisations can benefit from health and safety advice on key issues; obtain external evaluation of their health and safety management systems; get agreed interpretation on health and safety matters; show their commitment to health and safety; develop public and employee relations; and reduce administration.

6. Although a substantial part of the partner business should be local authority controlled, parts of the business subject to HSE enforcement can form part of the partnership agreement. Small and medium sized businesses with up to 250 employees can join these schemes provided their outlets or branches exist in two or more local authority areas.

7. Valuable partnerships can also be formed with trade and other associations representing the interests of a group of businesses. In these cases, of particular interest might be the quality of health and safety information issued; how health and safety legislation is interpreted for the benefit of the members; how health and safety is promoted; and the way in which communication and liaison takes place within the organisation and related bodies.

8. Local authorities can gain valuable experience of health and safety management auditing, often called Safety Management Reviews.

There are no set time-scales for completing the different stages of a partnership scheme, although once started the most benefit will be obtained from an agreed progress plan. Examples of partnerships seen to be successful by both partners include those developed by Coventry City Council and GEHE UK plc, the latter employing around 13,000 staff; Basingstoke and Deane Borough Council and The Automobile Association; and Wolverhampton MBC and Wolverhampton and Dudley Breweries plc. Partnerships such as these have produced Agreements of Intent.

A key component of any agreement is the commitment to a Safety

Management Review to evaluate the organisation's health and safety management systems. HELA organises courses on the necessary auditing skills. In a large organisation, a typical review might involve several local authority staff for one or two days per week over as much as six months, taking into account their other work obligations to their employing authority. A draft report follows, identifying the strengths and weaknesses of the health and safety management system and recommending improvements. Following approval of the draft report by the organisation, there will be a formal presentation to its senior management, following which it is expected to respond, preferably by preparing an action plan for dealing with the issues raised.

Some of the more common issues which have been raised include the following:

- Out of date policy statements which do not reflect the current company activities.

- Inadequate systems of work.

- The absence of current risk assessments and inadequate reviews of those that had previously been conducted.

- Varying levels of staff training and a lack of follow-up training.

- The absence of any reference to staff training in routine staff meetings.

- Little or no monitoring of health and safety performance.

- Lack of understanding of disciplinary proceedings associated with health and safety failings.

The kind of steps taken following Safety Management Reviews have included:

- Identification of training needs and the development of training programmes adapted to the needs of different staff.

- Full risk assessments and the development of new safety policies.

- Auditing of systems and procedures with review and revision where necessary.

- Action plans to address the findings of the assessments and reviews.

- Regular reviews of health and safety policies and work instructions.

- Increased employee involvement in health and safety, including the establishment of safety committees where they did not exist.

- Health and safety introduced as a standing item on company business reviews.

- Specific financial provision for health and safety in annual budget reviews.

- Improved systems for conveying health and safety information.

- Introduction of health and safety in induction programmes.

- Monitoring of performance standards and safety systems.

- Priority investigation of accidents and near misses.

- Regular auditing to ensure a sustainable programme of continuous improvement.

The HSE's Local Authority Unit is committed to supporting Lead Authority Partnership Schemes through providing information and guidance, helping to set up schemes, arranging training, commissioning research and providing impartial advice.

TRADING STANDARDS AUTHORITIES

Trading Standards Officers have an important role in dealing with health and safety issues, although that role is not always fully understood. Their main statutory functions fall under the Petroleum (Consolidation) Act 1928, the Dangerous Substances and Explosive Atmospheres Regulations 2002,[58] the Explosives Act 1875 as

---

[58]   S.I. 2002 No. 2776.

amended by the Explosives Act 1923, the Consumer Protection Act 1987 and the General Product Safety Regulations 1994.[59] Certain contraventions become prosecutable under the Health and Safety at Work etc. Act 1974 as "relevant statutory provisions" by virtue of Schedule 1 to the Act, e.g. in respect of explosives and petroleum-related offences. It is therefore important to understand how the work of trading standards officers and health and safety officers may be linked together.

In relation to petroleum, the main responsibilities are:

1. Licensing the keeping of petroleum spirit[60] and other substances.[61]

2. Control over the labelling of vessels containing petroleum spirit.[62]

3. Powers to test petroleum spirit[63] and to search for and seize it if it is being kept, sent, conveyed, or exposed or offered for sale in contravention of the Act.[64]

Enforcement is by County Councils, Unitary Authorities or Fire and Civil Defence Authorities. In the former case powers are usually given to trading standards officers and in the latter they may alternatively be given to fire officers.

In relation to explosives, the main responsibilities are:

1. Issuing consents for the establishment of new factories or magazines for gunpowder.[65]

2. Issuing licences for gunpowder stores.[66]

3. The registration of premises for the keeping of gunpowder,[67]

---

[59]   S.I. 1994 No. 2328.
[60]   Petroleum (Consolidation) Act 1928, ss.1-4.
[61]   *ibid.,* by orders made under s.19.
[62]   *ibid.,* s.5.
[63]   *ibid.,* s.17.
[64]   *ibid.,* s.18.
[65]   Explosives Act 1875, ss.6-8.
[66]   *ibid.,* ss.15 and 18.
[67]   *ibid.,* s.21.

which has to be kept under certain conditions.[68]

4.   Keeping registers of all store licences granted and premises registered.[69]

The majority of the other statutory controls over explosives are enforced by the Health and Safety Executive.

Explosives may be fireworks, or other types such as those used for blasting, e.g. in quarries and coal mines, guns and black powder used in historical battle re-enactments.

In the case of fireworks, there is often a dual and related role for trading standards and health and safety officers, the former being responsible for regulating the sale of fireworks to children who are under age, the latter being responsible for ensuring both the protection of employees who may store and sell fireworks, and protecting the public from the associated risks, e.g. the sale without proper handling instructions or with damaged wrapping which may introduce a greater risk during handling. In the case of explosives, the police also have responsibility for security at certain premises. Accordingly, there will be cases when it is desirable for all of the enforcing officers to liaise to ensure effective enforcement.

The Consumer Protection Act 1987 and the General Product Safety Regulations 1994[70] made under the European Communities Act 1972 provide one of the main areas of consumer protection involving health and, more particularly, safety. The general concept of the regulations is to ensure the safety of all products so far as is reasonably practicable. It may be useful to give a couple of examples of where statutory responsibilities can overlap in the same premises in order to understand the importance of the different enforcement officers working together when necessary:

1.   In farm shops selling goods for use at home and work, e.g. gloves, health and safety officers would ensure the goods

---

[68]   Explosives Act 1875, s.22.
[69]   *ibid.*, s.28.
[70]   S.I. 1994 No. 2328.

were safe for use at work and trading standards officers would ensure they met the standards required by consumer law.

2. In equipment hire shops, electrical wholesalers and builders merchants, trading standards officers would expect equipment to meet relevant manufacturing standards. Where the user (consumer) is a builder or similar tradesman, the officers would expect that person to be fully familiar with the safe operating requirements of any equipment hired. They would not necessarily require, therefore, the same detailed safety advice to be given in respect of equipment hired to a tradesman as might be required by a health and safety officer primarily concerned with protecting a member of the general public who might only hire the equipment occasionally.

In relation to consumer safety issues, the principal powers of trading standards officers are contained in the Consumer Protection Act 1987 and the General Product Safety Regulations 1994.[71]

It may also be useful to be aware of some potential conflicts/issues that may create confusion between health and safety and trading standards officers when it comes to enforcement:

•   The need to recognise the different skills and expertise of each group of officers as they relate to health and/or safety.

•   Ensuring the co-ordination of inspections/investigations at premises or activities giving rise to significant health and safety issues involving both regulators, e.g. petrol filling stations.

•   Failure to understand the statutory role and scope of their respective roles.

•   Use of different information databases for premises in the same local authority area.

•   Differences in enforcement philosophies. Trading standards officers often adopt an advisory role with enforcement as a last

---

[71] s.10(1)(a) and (2)(b) relating to published standards of safety for consumer goods may be a similar means of control to ss.3 and 6 of the Health and Safety at Work etc. Act 1974.

resort, whilst some health and safety officers may concentrate on a statutory enforcement approach as the preferred method of control.

Such issues can present difficulties, particularly where both services are in the same local authority department. In such cases, duty holders can find it especially confusing to deal with officers with different enforcement policies and philosophies. There are, however, steps that can be taken to avoid any confusion:

1. Staff training programmes should ensure that each group of officers is conversant with the basic legal duties and responsibilities of the other, particularly the extent of any overlap in their functions. This will enable officers to decide when it is appropriate to refer particular matters to the other group.

2. Where possible, information on premises should be on a common or compatible database so that whoever inspects them can update basic information for the benefit of both groups of enforcement officers.

3. Enforcement policies should either be consistent or, if not, each group should be aware of the differences.

4. Inspections of premises requiring a significant involvement from a trading standards and health and safety viewpoint could usefully be inspected jointly to ensure all safety issues are dealt with at the same time, e.g. petrol filling stations. Joint inspections are also useful when staff resources are short.

5. Regular management meetings involving staff from both areas of activity can help identify other issues which need to be addressed.

It will be seen that, although not primarily responsible for enforcing the Health and Safety at Work etc. Act 1974, trading standards, or consumer protection departments as they are often known, have an important role in the enforcement of health and safety issues through trading standards legislation. Consequently, liaison between trading standards and health and safety officers is particularly important where their functions may overlap. Following the abolition of the former Metropolitan County Councils by the Local Government Act 1985, in many areas trading standards functions are now part of the duties of Metropolitan District Councils. In some authorities they are in the same department as health and safety officers. They visit the same premises to carry out routine inspections, enforcement and sampling and it is important that they co-ordinate their activities on health and safety related issues. It has been known for two officers to arrive at the same premises to carry out an enforcement role without the prior knowledge of either of them. Examples include:

- Inspections at petrol filling stations where trading standards officers will carry out investigations relating to the licensing and storage of petroleum under the Petroleum (Consolidation) Act 1928 and Dangerous Substances and Explosive Atmospheres Regulations 2002, and health and safety staff will also investigate the management of safety precautions by virtue of their duties under the Health and Safety at Work etc. Act 1974.

- Visits to fireworks retailers where trading standards officers are checking on the sale of fireworks to minors. Health and safety staff will be ensuring that the employer has safe methods of storage (trading standards staff also have similar responsibilities under the Explosives Act 1875) and is not endangering members of the public through his activities; this might arguably include selling fireworks to someone incapable of using them correctly, e.g. a minor or someone with an obvious physical or mental disability. The Fireworks Act 2003 also makes provision for the Secretary of State to make regulations relating to the supply and use of fireworks. This was necessary as the existing powers under the Consumer Protection Act 1987 were limited to the supply, as opposed to the use of, consumer goods.

Liaison between these groups of officers is essential if limited resources are to be used effectively. Examples of good liaison include:

- In the case of an unco-operative employer, calling on other enforcement colleagues to see if action available to them, possibly including joint enforcement action, can produce a more positive response.

- Issuing by post self-assessment questionnaires on relevant health and safety or trading standards issues. These "postal inspections" will often produce information indicating premises where there is a poor understanding of an employer's statutory responsibilities and identify where officers need to carry out inspections and investigations. This usually involves inspecting a proportion of the premises from which there has been a return *but visiting all of those from which there has been no response.*

The trading standards "home authority" scheme which also forms the basis on which the health and safety Lead Authority Partnership Scheme was established works well but is reliant on the participants communicating relevant information to the lead authority. It does not result in all information about a particular premises or employer being made available to other authorities. In this respect some health and safety and trading standards officers feel that the partnership schemes could be improved by the development of a national database on which all information relating to premises covered by the schemes would be transferred. This would include data on routine inspections and investigations.

TRADES UNIONS

Whilst they do not have a statutory enforcement role, trades unions have a key role in implementing health and safety improvements through their safety officer representatives in the workplace and their wider role of training their members and influencing new legal developments. All of the major unions have taken a proactive role in responding to the governments' *Revitalising Health and Safety* strategy. It is therefore important for local authorities to be aware of what trades unions are doing to improve health and safety in the workplace, and where possible to work with them when appropriate or necessary.

For years the major trades unions, in conjunction with the TUC, have been campaigning against the decline in local authority resources devoted to health and safety, unfortunately with little effect. Accordingly, most trades unions provide services to their members and safety representatives which try to bridge the gap created by the shortfall in local authority activity. The TUC General Secretary has set out his views on the problems experienced with employers.[72] He rightly identifies four groups of employers:

1. The *criminal* – those who do not respond to any persuasion or argument.

2. The *clueless* – those who still do not know what are their health and safety responsibilities.

3. The *compliant* – those who only do what is absolutely necessary and do not really think about health and safety.

4. The *stars* – the committed employers who do more than just meet their basic legal obligations.

Local authority enforcement officers can probably relate to those views. The TUC prefers the partnership approach to securing improvements in health and safety and local authorities can potentially benefit from an understanding of the way in which trades unions are working, and by joint working arrangements. After all, they are all seeking to improve health and safety in the workplace. However, joint working with trades unions may not be a priority of all local authorities. A telling observation is made by USDAW:[73] "unfortunately we have found it difficult to build practical links between our safety representatives and local authority enforcement officers over the years. This may be partly due to a lack of knowledge about the presence of trades unions on the part of the local authority enforcement officers. It is also no doubt due to the lack of knowledge of the role that the local authority plays on the part of our members."

USDAW has started to address this issue by developing links

---

[72] In the Health and Safety Executive Northern Ireland Annual Lecture, University of Ulster, November 13, 2002.
[73] Private communication.

between union safety representatives, union officials and some local authorities. It has also produced a model letter intended to be sent by safety representatives to local authority enforcement officers which includes the following:

> "I understand that my functions under regulation 4(1) of the Safety Representatives and Safety Committee Regulations 1977 include:
>
> • Representing the employees in consultations at the workplace with inspectors from the enforcing authority.
>
> • Receiving information from the inspectors in accordance with section 28(8) of the Health and Safety at Work etc. Act 1974.
>
> I am therefore writing to let you know of my appointment. Please contact me when you are making future visits to the workplace."

HELA expect[74] that local authority inspectors, in carrying out their duties should:

(a) be supportive of safety/employee representatives in carrying out their functions as such representatives;

(b) make efforts to make contact with one or more safety/employee representatives when they visit workplaces;

(c) adopt a policy of openness in their dealings with safety/employee representatives, in accordance with the principles of openness and transparency, so far as the law allows them to do so.

This commitment to health and safety is reflected in other initiatives which may, or could in some cases, involve local authorities. Some examples are:

1. The TUC has asked unions to report regularly on:

   (a) what they are doing to raise awareness of the government's *Revitalising Health and Safety* strategy;

---

[74] See HELA LAC 73/2 (Rev), November 2000.

    (b) providing space in union journals for material on the *Revitalising* strategy;

    (c) discussing *Revitalising* targets with employers or their associations;

    (d) developing health and safety partnerships;

    (e) involvement at Ministerial or Commissioner level; and

    (f) conferences and visits to premises with partnerships for prevention purposes.

2.   The TUC launched a "Union Inspection Notice" in 2001. Loosely based on an Australian initiative, it is a voluntary system allowing safety representatives to draw to their employer's attention possible breaches of health and safety law. These notices are intended to be a last resort but offer the employer the opportunity to resolve issues without the union having to bring in enforcement officers.

3.   Amicus MSF has included sessions on *Revitalising* in its safety representative courses.

4.   UNISON, the GMB and TGWU have jointly pursued the adoption of health and safety targets for local authorities. These are now contained in the HSC's annual Business Plans.

5.   The NUT has issued guidance on targets and partnership initiatives to its Divisional Health and Safety Advisors.

6.   Amicus MSF, CWU, PCS, UNISON and USDAW are involved in a partnership on health and safety in call centres on Merseyside.

7.   UNIFI has a partnership with Barclays Bank providing for several full-time seconded union health and safety representatives covering the whole of the country.

8.   The TUC and its member unions have been involved in a Worker Safety Advisor pilot scheme designed to address the lack of employee involvement in health and safety in smaller

workplaces. The idea of the scheme was to provide small firms with trade union knowledge and expertise about worker involvement and to establish suitable working relationships. The scheme is reported to have produced a more positive approach to health and safety; workers being encouraged to suggest ways of improving the working environment; regular discussions on health and safety; improvements in compliance; better communications on health and safety; increased awareness of health and safety issues; and improved working relations in general.

9. UNISON and the Police Federation have signed a joint partnership agreement on an action plan to achieve health and safety targets with the Lancashire police force and the HSE.

10. USDAW is involved with local authorities, the dairy industry and retail and distribution employers in an HSE project on work-related transport in the milk industry in South West England.

11. UNISON and Newcastle City Council have produced agreements on a number of issues, including corporate guidance on liaison and co-operation on health and safety issues, and the approach to health and safety training for UNISON appointed safety representatives.

12. Most trades unions issue regular safety information sheets, safety guides, leaflets and posters, as well as other up-to-date information on health and safety developments.

Trades unions appear to be taking their health and safety responsibilities seriously and local authorities could well take advantage of this through a variety of possible partnership arrangements.

FIRE AUTHORITIES

Normally, the fire authority responsible for the enforcement of fire precautions in factories, offices, shops and railway premises[75] is,

---

[75] The law is contained in the Fire Precautions Act 1971, the Fire Precautions (Factories, Offices, Shops and Railway Premises) Order 1989, S.I. 1989 No. 76 and the Fire Precautions (Non-Certificated Factory, Office, Shop and Railway Premises) Regulations 1989, S.I. 1989 No. 78.

in England and Wales, the County Council and, in Scotland, the Islands or Regional Council. The Fire Precautions (Workplace) (Amendment) Regulations 2003[76] make the HSE the enforcing authority in respect of Part 2 of the Fire Precautions (Workplace) Regulations 1997,[77] i.e. for specific premises of a description contained in Part 1 of Schedule 1 to the Fire Certificates (Special Premises) Regulations 1976.[78] Under the latter regulations, in the case of factory, office, shop and railway premises within a defined list of special hazards, the enforcing and certificate granting authority is the Health and Safety Executive.[79]

In relation to fire protection, the Management of Health and Safety at Work Regulations 1999[80] require measures to be taken to ensure compliance with the Fire Precautions (Workplace) Regulations 1997.[81] The regulations variously require that, where necessary in order to safeguard the safety of employees in the case of fire:[82]

1. Workplaces must be provided with appropriate fire fighting equipment, fire detectors and alarms.

2. Non-automatic fire-fighting equipment that is provided must be easily accessible, simple to use and indicated by signs.

3. Where necessary, fire fighting measures must be provided, employees nominated to implement those measures and arrangements made for contacts with external emergency services, particularly regarding rescue work and firefighting.

4. Routes to emergency exits and the exits themselves must be kept clear at all times.

5. Where necessary, having regard to the features of the workplace, the nature of the activity, any hazard or other

---

[76] S.I. 2003 No. 2457.
[77] S.I. 1997 No. 1840.
[78] S.I. 1976 No. 2003.
[79] Acting under the Fire Certificates (Special Premises) Regulations 1976, S.I. 1976 No. 2003.
[80] S.I. 1999 No. 3242.
[81] S.I. 1997 No. 1840, as amended by S.I. 1999 No. 1877 and S.I. 2003 No. 2457.
[82] regs. 4 and 5.

relevant circumstances, the following particular requirements must be complied with:

(a) emergency routes and exits must lead as directly as possible to a place of safety;

(b) in the event of danger, it must be possible for employees to evacuate the workplace quickly and as safely as possible;

(c) the number, distribution and dimensions of emergency routes and exits shall be adequate, having regard to the use, equipment and dimensions of the workplace and the maximum number of persons that may be present there at any one time;

(d) emergency doors shall open in the direction of escape;

(e) sliding or revolving doors shall not be used for exits specifically intended as emergency exits;

(f) emergency doors shall not be so locked or fastened that they cannot be easily and immediately opened by any person who may require to use them in an emergency;

(g) emergency routes and exits must be indicated by signs; and

(h) emergency routes and exits requiring illumination shall be provided with emergency lighting of adequate intensity in the case of failure of their normal lighting.

Any equipment and devices provided in compliance with the regulations must be subject to a suitable system of maintenance and repair.[83]

Enforcement officers should ensure that they see suitable references to fire safety when examining a risk assessment. The above-mentioned requirements are included here as an *aide memoire*. If fire safety is not addressed by an employer, the matter should be referred to the relevant fire authority.

---

[83] reg. 6.

Fire authorities have similar powers to serve enforcement and prohibition notices under the Fire Precautions Act 1971 as local authorities and the HSE have under the Health and Safety at Work etc. Act 1974. Fire authorities are the principal enforcing authorities for premises where local authorities have a health and safety enforcement role. It is important that health and safety enforcement staff liaise closely with fire authorities on issues of fire safety to ensure that they are aware of what matters, if found, should be referred to those authorities. The fire authority contact is often someone at Divisional Officer level.

Chapter 3

STATUTORY DUTIES AND ENFORCEMENT

Although this book is primarily concerned with the enforcement of health and safety legislation by local authorities, in order to fully understand those responsibilities it is also necessary to understand the primary duties of employers, the self-employed and employees in complying with the Health and Safety at Work etc. Act 1974 and its associated regulations. Accordingly, this chapter also includes reference to those duties. It does not seek to cover every duty and responsibility, that would fill another book. It simply deals with the main requirements of employers in order to put local authority enforcement duties into context.

EMPLOYER AND EMPLOYEE RESPONSIBILITIES

General duties of employers to their employees

Section 2(1) of the Health and Safety at Work etc. Act 1974 states that it is the duty of every employer to ensure, *so far as is reasonably practicable,* the health, safety and welfare at work of all his employees.

The qualification of this duty by use of the phrase *so far as is reasonably practicable* places the burden of proving that what was done to comply with the law was all that was reasonably practicable on the defendant in a criminal prosecution.[1] Case law reveals a number of principles that are associated with this phrase:

1. The existence of a universal practice does not necessarily discharge the onus on an employer of proving[2] that it was not reasonably practicable to use some other safer method.

2. The proper test of what is "reasonably practicable" is not just whether the measures were physically or financially possible; the principle that the degree of risk has to be weighed against

---

[1] See s.40 of the Health and Safety at Work etc. Act 1974.
[2] Under s.40. See *Martin v. Boulton and Paul (Steel Construction) Ltd.* [1982] I.C.R. 366, DC.

the sacrifice involved must also be taken into account. If the sacrifice is disproportionately heavy in relation to the risk, then the measures are not "reasonably practicable".[3]

3. Section 2 is not intended to outlaw certain activities merely on the basis that they are dangerous.[4]

4. Any reasonable precaution which can be taken, however, must be taken.[5]

5. It is inappropriate to measure what is "reasonably practicable" by the content of safety regulations not yet in force.[6] Such regulations might, of course, be indicative of good practice.

6. A corporate employer, e.g. a limited company, is criminally liable, under section 2, where there is a failure to ensure the health, safety and welfare at work of any employee, unless all reasonable precautions have been taken by the corporate employer or on its behalf by its servants and/or agents. It is not necessary for the prosecution to show fault by head office personnel or senior management.[7]

7. Where safety devices are provided, e.g. by a specific regulation, the words "so far as is reasonably practicable" govern only the *capacity* of the safety devices and not the absolute obligation to provide such devices.[8]

---

[3] *West Bromwich Building Society v. Townsend* [1983] I.C.R. 257, DC, applying *Edwards v. National Coal Board* [1949] 1 K.B. 704.

[4] *Per* Turner J. in *Canterbury City Council v. Howletts and Port Lympne Estates, The Times,* December 13, 1996 – zoo's practice of allowing keepers to "bond" with tigers. In this case the local authority contended that during the cleaning process the tigers should be secured. The High Court affirmed the decision of an industrial tribunal to set aside the notice because the statutory intention was not to render all dangerous practices illegal.

[5] *Marshall v. Herbert* [1963] Crim. L.R. 506; *Garrett v. Boots the Chemists Ltd.* (1980) DC (unreported).

[6] *R. v. Birmingham City Council, ex p. Ferrero Ltd., The Times,* March 3, 1990.

[7] *R. v. Gateway Foodmarkets Ltd., The Times,* January 2, 1997, CA. It was held that the defendant could be liable under s.2(1) even though at senior management or head office level it had taken all reasonable precautions. There was no requirement for the prosecution to prove a failure at the level of the "directing mind" of the defendant company.

[8] *R. v. Rhone Poulenc-Rorer Ltd., The Times,* December 1, 1995, CA (obligations under reg. 36(2) of the Construction (Working Places) Regulations 1966, S.I. 1966 No. 94).

8.   The prosecution does not have to specify in the information anything more than a breach of the general duty in subsection (1). It does not have to refer to, nor specify breaches of the instances in, subsection (2).[9]

Section 2(2) states that, without prejudice to the generality of an employer's duty (under section 2(1)), the matters to which that duty extends include, in particular:

1.   The provision and maintenance of plant and systems of work that are, *so far as is reasonably practicable,* safe and without risks to health.

The duty under subsections (1) and (2)(a) is for employers to provide and maintain safe and risk free plant and systems of work for all their employees "at work" and not just for those working in the specific process for which the plant in question was made available. If an employer makes available unsafe plant, there is a breach of the duty, *even though the plant had not been and was not being used.*[10] Accordingly, it must be maintained and operated using a safe system, e.g. not overloaded or used for purposes for which it was not designed.

2.   Arrangements for ensuring, *so far as is reasonably practicable,* safety and absence of risks to health in connection with the use, handling, storage and transport of articles and substances.

This subsection relates only to the transport of goods, not people. Other provisions of this section would control risks associated with the transport of people.[11]

---

9   If it does refer to specific breaches and to more than one element of the sub-paras. of subs. (2), that does not render the information bad for duplicity: *Health and Safety Executive v. Spindle Select Ltd., The Times,* December 9, 1996, DC. *Cf.* The Scottish cases of *Cardle v. Carlaw Engineering (Glasgow)* 1991 S.C.C.R. 807 and *Carmichael v. Marks and Spencer* 1995 S.C.C.R. 781.

10   *Bolton Metropolitan Council v. Malrod Insulations Ltd.* [1993] I.C.R. 358, DC. For the nature of evidence required to sustain a prosecution, see *Tesco Stores Ltd. v. Seabridge, The Times,* April 29, 1988, DC (self-evident breach of the subsection).

11   e.g. s.2(2)(d), (e).

3. The provision of such information, instruction,[12] training and supervision as is necessary to ensure, *so far as is reasonably practicable,* the health and safety of his employees.

It has been held that the duty under subsection (2)(c) may require provision of information to non-employees working in connection with the employer's undertaking.[13] This duty is strengthened by the requirement on an inspector under section 28(8) to communicate to employees factual information obtained by him in the course of an inspection about matters affecting their health, safety and welfare. It also emphasises the need for employers to produce and maintain safe systems of work, to ensure that all employees are properly trained in advance of their use, and to monitor the use of the systems in practice.

4. *So far as is reasonably practicable,* as regards any place of work under the employer's control, the maintenance of it in a condition that is safe and without risks to health and the provision and maintenance of means of access to and egress from it that are safe and without such risks.[14]

5. The provision and maintenance of a working environment for his employees that is, *so far as is reasonably practicable,* safe, without risks to health, and adequate as regards facilities and arrangements for their welfare at work.[15]

Section 2(3) requires that, except in prescribed cases, it is the duty of every employer to prepare and, as often as may be appropriate, revise a written statement of his general policy with respect to the

---

12 "Instruction" has two meanings: to teach and to order, *Boyle v. Kodak* [1969] 2 All E.R. 439, and an employer who is to comply with his statutory duty may have to invoke disciplinary proceedings against an employee if he fails to follow rules intended to safeguard his health and safety.

13 *R. v. Swan Hunter Shipbuilders Ltd., The Times,* July 6, 1981, CA. This involved a failure to warn sub-contractors' employees of fire dangers. See also the Health and Safety Information for Employees Regulations 1989, S.I. 1989 No. 682.

14 With regard to disabled persons, see s.8A of the Chronically Sick and Disabled Persons Act 1970.

15 Breach of this duty by an employer, e.g. by subjecting an employee to the risk of inhaling other employees' cigarette smoke, though not directly actionable in damages, may nevertheless constitute constructive dismissal of the employee if he leaves the job as a result, giving rise to an award of compensation by an employment tribunal: *Waltons v. Morse and Dorrington* [1997] I.R.L.R. 488, EAT.

health and safety at work of his employees and the organisation and arrangements for the time being in force for carrying out that policy, and to bring the statement and any revision of it to the notice of all his employees.

The requirement is only to bring the statement *to the notice of employees,* not to give everyone a copy. General practice is to give all employees a copy of the statement, although details of the organisation and arrangements for carrying it out, including the identification of personnel with specific safety responsibilities, are often contained in working documents, the whereabouts of which are generally made known to all employees and are usually referred to in training sessions.[16] There is an exemption from the duty to provide a health and safety policy where any employer carries on an undertaking in which for the time being he employs less than five employees.[17]

The main components of a statement of health and safety policy should be:

1. A statement of intent describing the organisation's philosophy on the management of health and safety. This should reflect the duties contained in section 2 of the Act and set out the principle objectives for complying with the law.

2. The organisational structure relating to health and safety. This should include the hierarchy of responsibility and accountability at all levels.

3. The detailed arrangements, including the systems and procedures for implementing and monitoring health and safety performance against the objectives contained in the statement

---

[16] The HSC has issued advice to employers on the production of health and safety policies, see "Enforcement Policy Statement", HSC 15, January 2002, HSE.

[17] Employers' Health and Safety Policy Statements (Exception) Regulations 1975, S.I. 1975 No. 1584. In the case of *Osborne v. Bill Taylor of Huyton Ltd.* [1982] I.R.L.R. 17, the defendant operated 31 betting shops and argued that one of these, which was the subject of an improvement notice requiring a safety policy, did not require one as that particular shop employed less than five people. The contention that no safety policy was required was upheld. In practice, however, such multiple premises organisations will generally produce an overall policy applicable to their business.

of intent. This should include the provision of information, instruction and training, risk assessment procedures, reporting of accidents, and investigation and recording procedures.

It is quite common for policy statements to contain more detailed secondary statements about individual policies which may be of particular relevance to the organisation, e.g. training, staff security, smoking and health surveillance. It is also useful to include details of all relevant documents that are available for staff perusal, together with their location.

Section 2(6) places a duty on every employer to consult any safety representatives[18] that have been appointed with a view to the making and maintenance of arrangements which will enable him and his employees to co-operate effectively in promoting and developing measures to ensure the health and safety at work of the employees, and in checking the effectiveness of such measures. In prescribed cases it is also the duty of every employer, if requested to do so by a safety representative, to establish, in accordance with the regulations,[19] a safety committee having the function of keeping under review the measures taken to ensure the health and safety at work of his employees and such other functions as may be prescribed.

Employers also have to consult their employees, where they are not represented by safety representatives,[20] in good time on matters relating to their health and safety at work and, in particular, on certain specific issues which include the introduction of any measure at the workplace which may substantially affect the health and safety of those employees. The consultation with employees must be either with the employees directly or, in respect of any group of employees, with representatives elected by the group. In most other respects the regulations contain similar provisions to the 1977 regulations.

---

[18]   Appointed by virtue of regulations made under subs.(4). These are the Safety Representatives and Safety Committees Regulations 1977, S.I. 1977 No. 500. For a more detailed explanation of these regulations, see pp.84 and 85.
[19]   *ibid.,* s.2(7).
[20]   Health and Safety (Consultation with Employees) Regulations 1996, S.I. 1996 No. 1513, as variously amended following the decision in *E.C. Commission v. United Kingdom* [1995] 1 C.M.L.R. 345; [1994] I.C.R. 664, E.C.J.

General duties of employers and self-employed to persons other than their employees

Section 3(1) imposes a duty on every employer to conduct his undertaking in such a way as to ensure, *so far as is reasonably practicable,* that persons not in his employment who may be affected by it are not exposed to risks to their health and safety.

As with section 2, a number of principles arise from case law which may be summarised as follows:

1. The subsection imposes absolute criminal liability, subject only to the defence of reasonable practicability, this defence relating only to measures necessary to avert the risk.[21] A corporate defendant cannot escape criminal liability simply by showing that senior management had taken reasonable care to delegate fulfilment of the duty under this subsection.[22]

2. The general duty includes a duty in appropriate cases to give information to non-employees, even though subsection (3) makes specific provision for this in prescribed cases.[23]

3. If the risk of exposure to danger is present, it is immaterial that the risk did not actually materialise.[24]

4. Duties are imposed on employers in relation to risks to members of the public,[25] including injured guests in hotels, guest houses, etc.

5. The conduct of an undertaking is not confined to situations where the undertaking is being actively carried on but is more general in its scope, e.g. a cleaning company was liable to a non-employee who was electrocuted whilst using its defective cleaning machine;[26] and there was a liability to a young person killed on a motor-cycle training course.[27]

---

[21] *R. v. British Steel plc* [1995] I.C.R. 587, CA.
[22] *ibid.*
[23] *R. v. Swan Hunter Shipbuilders Ltd.* [1981] I.C.R. 831, CA.
[24] *R. v. Board of Trustees of the Science Museum* [1993] 1 W.L.R. 1171, CA (risk of legionella infection; the duty was broken even if there was no inhalation or proof of the existence of the bacteria).
[25] *ibid.*
[26] *R. v. Mara* [1987] 1 W.L.R. 87, CA.
[27] *Mid-Suffolk D.C. v. Parkside (Care) Ltd.* (11.3.96) C.L.Y. 2996.

6. An employer does not "conduct his undertaking" if he employs an independent contractor to do the relevant work, *provided the employer neither exercises any control over the work, nor is under any duty to do so.*[28]

7. An employer may be able to argue the *reasonably practicable* defence in a case where his employee has been negligent and exposed a third party to a health and safety risk, provided he can show that he has fully trained and supervised his employee and provided him with safe, suitable equipment.[29] The employee may, of course, be liable.

8. The prosecution must give particulars in the summons, etc. of how the assessment of risk was inadequate and the system of work unsafe.[30]

This subsection is particularly relevant in cases where two or more employers are working at the same location and their activities may expose all employees to a risk to their health and safety, e.g. building sites, installation or maintenance work at commercial premises, contract catering or cleaning activities. In such cases it may be necessary to ensure safe working arrangements through contractual arrangements containing details of the systems required to ensure safety. Contractual arrangements, however, only work well if all parties and their employees are made fully aware of what is required. Issuing copies of a safety policy and the contract will rarely be enough: instruction and training will invariably be needed as well. If a subcontractor considers that the working arrangements under which he has to operate are unsafe, he has a duty either to ensure that things are put right or to terminate the contract. The key to joint working is co-ordination at all stages to ensure that health and safety is not compromised.

---

[28] *R.M.C. Roadstone Products Ltd. v. Jester* [1994] I.C.R. 456, DC, but see also *R. v. Associated Octel Ltd.* [1994] I.R.L.R. 540, *The Times,* August 3, 1994, CA, where it was held that a company was criminally liable to the employee of a cleaning contractor (carrying out annual maintenance) for a breach of s.3(1). Even though the plant itself was shut down, the company could still be said to be conducting its undertaking. The question of control by the company was very relevant to what was "reasonably practicable".

[29] *R. v. Nelson Group Services (Maintenance) Ltd.* [1998] 4 All E.R. 331, CA.

[30] *Heeremac VOF v. Munro* 1999 S.L.T. 492, HCJ Appeal (Scot.).

Section 3(2) requires every self-employed person to conduct his undertaking in such a way as to ensure, *so far as is reasonably practicable,* that he and other persons (not being his employees) who may be affected by it are not exposed to risks to their health and safety.

An example of this obligation is the case of a butcher who appealed against an improvement notice requiring him to wear a chain mail apron to protect himself when boning meat. The appeal was dismissed.[31]

Section 3(3) requires that, in prescribed cases and circumstances, every employer and self-employed person must give to persons who are not his employees but who may be affected by the way in which he carries on his undertaking, prescribed information about such aspects of the way in which he conducts his undertaking as might affect their health and safety. No regulations have been made in relation to this but an employer's duty under subsection (1) not to expose non-employees to risks to their health and safety may extend to a duty to give information and instruction as to such risks.[32]

General duties of persons concerned with premises to persons other than their employees

Section 4(1) of the Health and Safety at Work etc. Act 1974 imposes on persons in control of premises duties in relation to those who:

(a)   are not their employees; but

(b)   use non-domestic premises made available to them as a place of work or as a place where they may use plant or substances provided for their use there,

and applies to those premises made available and other non-domestic premises used in connection with them.

---

[31]   *Jones v. Fishwick* (1989) (COIT 10156/89).

[32]   *Carmichael v. Rosehall Engineering Works* [1983] I.R.L.R. 480 (High Court of Justiciary – Scotland: alleged failure to inform a person seconded to employers as part of a Manpower Services Commission Course).

This includes the common parts of a block of flats, e.g. the lifts and electrical installations and workmen repairing, etc. such parts.[33] It also includes a children's indoor play centre[34] and an open work site.[35]

Section 4(2) places a duty on each person who has, to any extent, control of premises[36] or of the means of access and egress, or of any plant or substance in such premises, to take such measures *as it is reasonable* for a person in his position to take to ensure, *so far as is reasonably practicable,* that the premises, all means of access and egress available for use by persons using the premises, and any plant or substance in the premises or provided for use there, is or are safe and without risks to health. Where a person has, by virtue of any contract or tenancy, an obligation of any extent relating to the maintenance or repair of any premises, the means of access or egress; or the safety of or absence of risks to health arising from plant or substances in any such premises, then that person is one to which this subsection applies.[37]

This subsection places an absolute duty, subject only to the limited qualification "so far as is reasonably practicable". However, if the premises are not a reasonably foreseeable cause of danger to persons using them in a manner or in circumstances which might reasonably be expected to occur, it is not reasonable to require any further measures to be taken against unknown and unexpected events.[38]

The position relating to domestic employment
Nothing in Part 1 of the Act applies "in relation to a person by reason only that he employs another, or is himself employed, as a

---

[33] *Westminster City Council v. Select Management Ltd.* [1984] 1 All E.R. 994 affirmed by CA [1985] 1 W.L.R. 576 (see also the definition of "domestic premises" in section 53(1)).

[34] *Moualem v. Carlisle City Council, The Times,* July 8, 1994, DC.

[35] *Geotechnics v. Robbins* (CO 470/95) [1995] C.L.Y. (electricity discharge from an overhead cable).

[36] A wide meaning was given to this phrase in *T. Kilroe and Sons v. Gower* [1983] Crim. L.R. 548, Liverpool Crown Court. Subsection (4) also qualifies the reference to a person having control, being a person having control of the premises or matter in connection with the carrying on by him of a trade, business or other undertaking whether for profit or not.

[37] s.4(3)(a), (b).

[38] *Mailer v. Austin Rover Group Ltd.* [1989] 2 All E.R. 1087, HL.

domestic servant in a private household".[39] The provisions of this section do not, however, prevent enforcement authorities from using the Act in appropriate cases involving domestic premises. In particular, the Act may need to be enforced in cases where accidents have occurred in connection with installation, construction and maintenance work by commercial organisations, e.g. installation of gas, electricity or water supplies; roofing and brickwork repairs; building extension and drainage works. Whilst the Act will not apply to DIY activities, if such activities were undertaken as part of construction or refurbishment work on domestic property as a commercial activity, e.g. converting a house into flats for rent, then a statutory liability may apply in the case of an accident occurring during that work, or in the case of a subsequent incident resulting from the manner in which the works were carried out.

General duty of persons in control of certain premises in relation to harmful emissions into the atmosphere

Section 5(1) makes it the duty of the person having control of any prescribed class of premises to use the *best practicable means* for preventing the emission into the atmosphere, from the premises, of noxious or offensive substances[40] and for rendering harmless and inoffensive such substances as may be emitted.

The whole of section 5 is repealed by Schedule 16 to the Environmental Protection Act 1990, subject to Commencement Orders, section 162 of that Act, and the matters mentioned in section 162. In the meantime, the Environment Agency and the Scottish Environment Protection Agency are the enforcing authorities for the purposes of the Health and Safety at Work etc. Act 1974, section 5.[41]

General duties of manufacturers, etc. as regards articles and substances for use at work

Section 6 of the Act is enforced by the Health and Safety

---

[39] s.51.
[40] As specified in Sch. 2 to the Health and Safety (Emissions into the Atmosphere) Regulations 1983, S.I. 1983 No. 943, as amended by S.I. 1989 No. 319.
[41] By virtue of the Environment Act 1995, Sch. 22, para. 30.

Executive[42] but it is important to be aware of its provisions in order to know if its requirements are being met in those premises subject to local authority enforcement. If not, then relevant details should be passed to the Health and Safety Executive for enforcement action.

Section 6(1) requires any person who designs, manufactures, imports or supplies any article for use at work or any article of fairground equipment:

(a) to ensure, *so far as is reasonably practicable,* that the article is so designed and constructed that it will be safe and without risks to health at all times when it is being set, used, cleaned or maintained by a person at work;

(b) to carry out or arrange for the carrying out of such testing and examination as may be necessary for the performance of the duty imposed on him by the preceding paragraph;

(c) to take such steps as are necessary to secure that persons supplied by that person with the article are provided with adequate information about the use for which the article is designed or has been tested and about any conditions necessary to ensure that it will be safe and without risks to health at all such times as are mentioned in paragraph (a) above and when it is being dismantled or disposed of; and

(d) to take such steps as are necessary to secure, *so far as is reasonably practicable,* that persons so supplied are provided with all such revisions of information provided to them by virtue of the preceding paragraph as are necessary by reason of its becoming known that anything gives rise to a serious risk to health or safety.

This section was amended by the Consumer Protection Act 1987 which places a similar duty on the sellers of dangerous goods. The effect of the amendments was stated in a Health and Safety Bulletin:[43]

---

[42] By virtue of reg. 4(4)(a), Health and Safety (Enforcing Authority) Regulations 1998, S.I. 1998 No. 494.
[43] E2: 88 (19/1/88).

"The main effect is to ensure that designers, manufacturers, suppliers and importers are in no doubt that they must take account of reasonably foreseeable circumstances in which their products might be used, maintained, stored, etc., and provide relevant health and safety information with their product rather than simply make it available..."

Comparisons can be made between this section and:

(a) the terms implied on the sale, etc. of goods and the supply of services under the provisions of the Sale of Goods Act 1979 and the Sale and Supply of Goods Act 1994;

(b) the provisions of the Consumer Protection Act 1987 and associated regulations;[44]

(c) regulations implementing EEC Directives on injurious substances;[45]

(d) regulations relating to the packaging and labelling of dangerous substances;[46] and

(e) regulations requiring a European Communities Certificate for different types of plant, machinery and equipment.[47]

It is important to recognise that there is a considerable amount of similar or overlapping legislation enforced by different enforcement bodies that will often require close co-ordination to ensure effective, and often joint, enforcement.

Duties of employees at work

Section 7 places a duty on every employee while at work:

(a) to take reasonable care for the health and safety of himself and of other persons who may be affected by his acts or omissions at work; and

---

[44] e.g. Dangerous Substances and Preparations (Safety) (Consolidation) Regulations 1994, S.I. 1994 No. 2844, as variously amended.

[45] e.g. Good Laboratory Practice Regulations 1999, S.I. 1999 No. 3106.

[46] Classification, Packaging and Labelling of Dangerous Substances Regulations 1984, S.I. 1984 No. 1244.

[47] e.g. Construction Plant and Equipment (Harmonisation of Noise Emission Standards) Regulations 1988, S.I. 1988 No. 361 as variously amended.

(b) as regards any duty or requirement imposed on his employer or any other person by or under any of the relevant statutory provisions, to co-operate with him so far as is necessary to enable that duty or requirement to be performed or complied with.

There is little case law on this requirement but it could be applied to an employee driving a vehicle on an employer's private road where he caused an accident through careless driving, provided he was "at work".[48]

The Management of Health and Safety at Work Regulations 1999[49] place further duties on employees. In particular, regulation 14(1) requires that every employee using any machinery, equipment, dangerous substance, transport equipment, means of production or safety device provided by his employer in accordance with any training and information provided by his employer, must inform his employer or other employee with specific health and safety responsibilities, of any work situation which would reasonably be considered to represent a *serious and immediate danger to health and safety;* and of any other matter which would reasonably be considered to represent a *shortcoming in the employer's protection arrangements for health and safety.*[50]

Duty not to interfere with or misuse things provided pursuant to certain provisions

Section 8 requires that no person shall intentionally or recklessly interfere with or misuse anything provided in the interests of health, safety or welfare in pursuance of any of the relevant statutory provisions.

In proceedings for an offence, the prosecution has to establish that the accused intended the interference, and "intention" has been interpreted as being synonymous with "aim". "Intention" has been defined as "a decision to bring about, insofar as it lies within the accused's power, [a particular consequence], no matter whether

---

[48] *Coult v. Szuba* [1982] I.C.R. 380, DC.
[49] S.I. 1999 No. 3242.
[50] reg. 14(2)(a) and (b).

the accused desired that consequence of his act or not."[51] The definition of the word "reckless" has been considered in a number of cases, none of which relate specifically to health and safety legislation. In one case[52] the meaning was explained as "In my opinion, a person charged with an offence is reckless ... if (1) he does an act which in fact creates an obvious risk ... and (2) when he does the act he either has not given any thought to the possibility of there being any such risk or has recognised that there was some risk involved and has none the less gone on to do it." However, in another case involving different legislation[53] it was held that the prosecution had to prove that either the defendant intended, or that he actually foresaw, his act would cause harm. In yet another case, recklessness was equated with gross negligence.[54] Any prosecution under section 8 would therefore be interesting!

Proceedings for an offence could be taken against members of the public as well as employers or employees, although there does not appear to be any case law on this issue. The limitation in the section is on the conduct of the person, not the class of person. Examples of offences might include someone removing machinery guarding, the bypassing of an electrical safety cut-out device, and an employee or member of the public deliberately removing a safety warning sign or emergency exit sign. Whether there would be liability for well-intentioned but potentially dangerous acts, e.g. switching off an excessively noisy fire alarm or a smoke alarm thought to be unduly sensitive, is questionable.

The liability is only in relation to something provided to comply with safety law. Accordingly, dangerous practical jokes, e.g. leaving water taps on causing flooding, or causing injury by removing a chair as someone is about to sit down, would not necessarily result in liability under this section. However, such practices would almost certainly create a liability under section 7 on the individual foolish enough to engage in such horseplay.

---

[51] *R. v. Mohan* [1976] Q.B. 1.
[52] *R. v. Caldwell* [1982] A.C. 341, in relation to the Criminal Damage Act 1971, s.1.
[53] *R. v. Savage* [1991] 3 W.L.R. 914, relating to the Offences Against the Person Act 1861, s.20.
[54] *R. v. Holloway* [1993] 3 W.L.R. 927.

Duty not to charge employees for things done or provided pursuant to certain specific enactments

Section 9 prohibits an employer from charging any employee in respect of anything done or provided in pursuance of any specific requirement of the relevant statutory provisions.

DUTIES UNDER THE MANAGEMENT OF HEALTH AND SAFETY AT WORK REGULATIONS 1999[55]

The main purpose of these regulations is to impose requirements on employers which supplement the general duties contained in the 1974 Act, primarily in relation to their own employees. It must also be borne in mind that the enforcing authority is also an employer. As such it does not only have the responsibility for enforcing the law against employers but also complying with it. This means that it must also assess the implications of these regulations and other aspects of health and safety law in respect of the duties it expects of its health and safety inspectors. Prosecutions will often include both a breach of the regulations and a breach of the duty of care under the Act.

The principal duty is contained in regulation 3. Regulation 3(1) requires every employer to make a suitable and sufficient assessment of:

(a) the risks to the health and safety of his employees to which they are exposed whilst they are at work; and

(b) the risks to the health and safety of persons not in his employment arising out of or in connection with the conduct by him of his undertaking,

for the purpose of identifying the measures he needs to take to comply with the requirements and prohibitions imposed upon him by or under the relevant statutory provisions. The latter requirement will include such people as contractors, members of the public and guests or visitors to the place of employment.

---

[55] S.I. 1999 No. 3242. The regulations are accompanied by an HSC Approved Code of Practice and Guidance. For more detail on how to conduct a risk assessment, see Chapter 5, pp.234-241.

Regulation 3(2) imposes virtually the same requirements on the self-employed.

Regulation 3(3) requires the employer or self-employed person who carried out the assessment to review it if there is reason to suspect that it is no longer valid or there has been a significant change in the matters to which it relates and, if necessary, the assessment must then be modified.

Regulation 3(4) and (5) requires that no young person can be employed unless a risk assessment has been conducted and that assessment must take account, amongst other things, of the inexperience, lack of awareness of risks and immaturity of young persons.

Regulation 3(6) requires that, where an employer employs five or more employees, he shall record:

(a) the significant findings of his assessment; and

(b) any group of his employees identified by it as being significantly at risk.

It is important to be aware that there are many regulations that require risks to be assessed and certain risks are covered by more than one of these regulations. Only examination of the individual regulations will identify the differences in the risk assessment requirements. The risk assessment provisions most commonly encountered are contained in the following regulations:

• Management of Health and Safety at Work Regulations 1999.[56]

• Manual Handling Operations Regulations 1992.[57]

• Personal Protective Equipment at Work Regulations 1992.[58]

• Health and Safety (Display Screen Equipment) Regulations 1992.[59]

---

[56] S.I. 1999 No. 3242.
[57] S.I. 1992 No. 2793.
[58] S.I. 1992 No. 2966.
[59] S.I. 1992 No. 2792.

- Noise at Work Regulations 1989.[60]

- Control of Substances Hazardous to Health Regulations 2002.[61]

- Control of Asbestos at Work Regulations 2002.[62]

- Control of Lead at Work Regulations 2002.[63]

There are a lot of features common to risk assessments associated with these regulations.[64] Various approaches to a risk assessment may be adopted, including:

(a) an assessment of each activity or process likely to cause injury;

(b) an assessment of each type of plant, equipment, transport or material;

(c) an assessment of each different location, e.g. office, department, section or building.

Principles of risk prevention

The Management of Health and Safety at Work Regulations 1999[65] require that, where an employer implements any preventive and protective measures, he must do so on the basis of the principles specified in Schedule 1 to the regulations, as follows:

SCHEDULE 1
GENERAL PRINCIPLES OF PREVENTION
(Giving effect to Art 6(2) of Council Directive 89/391/EEC)

(a) avoiding risks;

(b) evaluating the risks which cannot be avoided;

(c) combating the risks at source;

---

[60] S.I. 1989 No. 1790.
[61] S.I. 2002 No. 2677.
[62] S.I. 2002 No. 2675.
[63] S.I. 2002 No. 2676.
[64] See *A guide to risk assessment requirements,* INDG 218, March 2002, HSE.
[65] reg. 4.

(d) adapting the work to the individual, especially as regards the design of workplaces, the choice of work equipment and production methods, with a view, in particular, to alleviating monotonous work and work at a predetermined work-rate and to reducing their effect on health;

(e) adapting to technical progress;

(f) replacing the dangerous by the non-dangerous or the less dangerous;

(g) developing a coherent overall prevention policy which covers technology, organisation of work, working conditions, social relationships and the influence of factors relating to the working environment;

(h) giving collective protective measures priority over individual protective measures; and

(i) giving appropriate instructions to employees.

Health surveillance

Every employer has a duty to ensure that his employees are provided with such health surveillance as is appropriate, having regard to the risks to their health and safety which are identified by the risk assessment.[66] The assessment will identify circumstances in which health surveillance is required by specific regulations, that surveillance enabling any adverse health effects to be detected, control measures taken and subsequent monitoring carried out.

Health and safety assistance

Every employer has a duty, subject to detailed conditions, to appoint one or more competent persons to assist him in complying with his obligations under the relevant statutory provisions and his fire precautions responsibilities.[67] Anyone appointed for this purpose must be competent and provided with adequate information and support. It is preferable for this person to be an employee, but if no competent person exists it may be necessary to appoint an

---

[66] reg. 6.
[67] reg. 7.

external expert, or in some cases an employee may be supported by an external expert. However competent the person appointed, such appointments do not absolve the employer from his responsibilities under the Health and Safety at Work etc. Act 1974.

Procedures for serious and imminent danger and for dangerous areas

Every employer must have in place appropriate procedures to be followed in the event of serious and imminent danger to persons working in his undertaking, must nominate a competent person to implement those procedures in relation to evacuation of the premises and must restrict access to dangerous areas only to those who have been adequately instructed.[68] The risk assessment should identify foreseeable risks that require the implementation of these procedures. It should also take account of the presence of any other employers or self-employed people and in such cases emergency procedures should be co-ordinated. Employers must arrange any necessary contacts with external emergency and other services that may be required.[69]

In the event that a risk assessment has not been carried out and any such dangers identified, the employer runs the risk of having improvement and/or prohibition notice(s) being issued by the enforcing authority to rectify the situation and any serious risks an inspector may identify.

Information for employees

Having completed a risk assessment and produced the necessary safety procedures, it is the duty of every employer to provide for his employees comprehensive and relevant information on the risks identified, the preventative and protective measures, the safety procedures and the person nominated with regard to evacuation procedures. The relevant information on risks and preventive measures will be limited to what employees need to know to ensure their own health and safety and not to put others at risk. It should be provided in a language and form understandable to those who need it.[70]

---

[68] reg. 8.
[69] reg. 9.
[70] reg. 10.

Where two or more employers share a work place, they are obliged to co-operate with one another in addressing the requirements of the regulations and to take reasonable steps to inform the other employees of the risks arising from their undertaking.[71] There are similar duties to provide information to external employers whose employees are working in a business, e.g. maintenance contractors.[72]

Capabilities and training

Every employer has a duty, in entrusting tasks to his employees, to take account of their capabilities as regards their health and safety. Employees have to be provided with adequate health and safety training, both on recruitment and on being exposed to new or increased risks because of changes in their duties or changes in the work or systems of work.[73] The training has to be repeated periodically where appropriate, adapted to take account of new or changed risks to health and take place in working hours.

Employees should not be allocated jobs beyond their capabilities and the level of their training, knowledge and experience. Employers should review their employees' capabilities to carry out their work as necessary, providing any additional training as required. This duty equally applies to health and safety inspectors who may require additional training to equip them to deal with new activities and risks associated with new types of activities which may be allocated to them, whether as a consequence of changes in the enforcing authority regulations or their employing authority decisions to change their duties and responsibilities.

An interesting initiative for organisations who wish to ensure their employees have basic health and safety awareness training is the "passport training scheme".[74] The "passports" are quite simply documents showing that a worker has up-to-date basic health, safety and environmental awareness training and possibly training related to other specific health and safety subjects. The "passport" is a simple way for workers moving from one type of employment

---

[71]   reg. 11.
[72]   reg. 11.
[73]   reg. 13.
[74]   Described in *Passport schemes for health, safety and the environment – a good practice guide,* October 2003, HSE.

to another, or who work in more than one type of employment, to show employers they have basic health and safety training.

DUTIES UNDER THE SAFETY REPRESENTATIVES AND SAFETY COMMITTEES REGULATIONS 1977[75]

Employer's duty to consult employees

Every employer has a duty to consult any appointed safety representative on measures associated with health and safety at work,[76] and in prescribed cases the employer must, if requested to do so by a safety representative, set up a safety committee.[77] He does not have to do so unless requested. Every employer must provide such facilities and assistance as safety representatives may reasonably require for the purpose of carrying out their functions.[78]

The safety representative, being a person representing the interests of employees, although an employee, is not a member of the employers' management team and has functions different from a safety manager. The safety manager is employed to advise the employer on safety matters and to deal with safety issues and will be considered a "competent person".[79] The safety representative's role is primarily an internal policing role, carrying out his own assessments of health and safety in monitoring management performance. He may, therefore, be in conflict rather than co-operating with management. However, the main trades unions prefer to encourage their safety representatives to co-operate both with employers and the enforcing authorities in improving health and safety. In fact, because of their intimate knowledge of the practices operating in their own workplaces, safety representatives are a useful source of information for inspectors investigating health and safety matters in the workplace. In many large workplaces, there may be several safety representatives all of whom should work together on such matters. USDAW and other

---

[75] S.I. 1977 No. 500.
[76] Health and Safety at Work etc. Act 1974, s.2(6).
[77] *ibid.,* s.2(7).
[78] reg. 4A, Safety Representatives and Safety Committees Regulations 1977, S.I. 1977 No. 500.
[79] For the purpose of reg. 6 of the Management of Health and Safety at Work Regulations 1999.

trades unions encourage the representatives to make contact and work with inspectors, and USDAW has a model letter which it suggests its representatives send to local enforcement officers encouraging contact and working together on health and safety matters.

Although the TUC has reported that workplaces with safety representatives and safety committees have significantly better accident records than those with no consultation mechanism, recording up to 50% fewer injuries, the majority of the workforce are not trades union members, or work in places where there is no union recognition. One option that has been considered to improve representation in small employers that may be worth future consideration by employers is the use of roving health and safety advisors.[80]

DUTIES UNDER THE USE OF WORK EQUIPMENT REGULATIONS 1998[81]

These regulations require that every employer shall ensure that work equipment[82] is so constructed or adapted as to be suitable for the purpose for which it is used or provided.[83] In selecting work equipment, employers must have regard to the working conditions and to the risks to the health and safety of persons which exist in the premises or undertaking in which that work equipment is to be used, and any additional risk imposed by the use of that work equipment.[84] Employers must ensure that work equipment is used only for operations for which, and under conditions for which, it is suitable.

The regulations also specify requirements as to the maintenance[85] and inspection,[86] and where there are specific risks to health and safety restricting the use, repair, modification, maintenance and

---

[80] For details see "Health and Safety Commission: Pilots to explore effectiveness of workers' safety advisors", interim findings, March 2003, York Consulting.

[81] S.I. 1998 No. 2306 as variously amended.

[82] Defined in reg. 2.

[83] reg. 4(1).

[84] reg. 4(2).

[85] reg. 5.

[86] reg. 6.

servicing to properly trained people.[87] Adequate health and safety information and instructions and training must also be provided for those who use the equipment.[88] The regulations make particular provision for protection against specified hazards[89] and the operation of particular controls,[90] controls over mobile work equipment[91] and power presses.[92]

The definition of "work equipment" covers any machinery, appliance, apparatus, tool or installation for use at work, and there are, therefore, a number of situations where the regulations may have direct implications for the health and safety staff carrying out routine inspections or investigatory work, including:

(a) operation of mobile laboratories;

(b) gas detection/breathing apparatus;

(c) use of portable scaffolding and ladders;

(d) use of sampling equipment.

This is yet a further example of how health and safety legislation affects the role of local authorities, both as enforcers and duty holders.

DUTIES UNDER THE WORKPLACE (HEALTH, SAFETY AND WELFARE) REGULATIONS 1992[93]

These regulations apply to a wide range of workplaces[94] and aim to meet the health, safety and welfare needs of each member of the workforce, which may include people with disabilities. A number of the regulations require things to be suitable[95] *for any person* in respect of whom things are done or provided. This makes it clear that traffic routes, facilities and workstations used by people with disabilities should be suitable for them to use.

---

[87] reg. 7.
[88] regs. 8, 9.
[89] reg. 12.
[90] regs. 13–24.
[91] Part III.
[92] Part IV.
[93] S.I. 1992 No. 3004.
[94] Defined in reg. 2(1).
[95] reg. 2(3).

The regulations apply to factories, offices and shops, but also to schools, hospitals, hotels and places of entertainment. The term workplace applies to the common parts of shared buildings, private roads and paths on industrial estates and business parks, and temporary work sites (excluding construction sites).

Employers have a general duty under section 2 of the Health and Safety at Work etc. Act 1974 to ensure, so far as is reasonably practicable, the health, safety and welfare of their employees at work, and these regulations expand on those duties. They are intended to protect the health and safety of everyone in the workplace and to ensure that adequate welfare facilities are provided for people at work. Employers must ensure that workplaces under their control comply with the regulations.

In particular, an employer must ensure that every workplace, modification or conversion which is under his control and where any of his employees works complies with any of the relevant requirements of the regulations.[96] The workplace and the equipment, devices and work systems must be maintained (including cleaning) in an efficient state, in efficient working order and in good repair.[97] It is no defence, for example that:

(a) the cause of failure of a lift could not be discovered;[98]

(b) the maintenance of a lift had been delegated to an independent contractor;[99]

(c) the cause of an unexpected fall of a curtain rail around a bed was unknown.[100]

There must also be suitable systems of maintenance[101] and at the very least such systems might need to include:

---

[96] reg. 1(4).
[97] reg. 5(1). The maintenance requirement is an absolute duty and not limited to what is reasonably practicable.
[98] *Galashiels Gas Co. Ltd. v. O'Donnell* [1949] A.C. 275, HL.
[99] *Malcolm v. Metropolitan Police Commissioner* [1999] C.L.Y. 1494 & 2880.
[100] *McLaughlin v. East Midlands NHS Trust,* 2000 Rep. L.R. 87, OH (Scot.). *Cf. McNaughton v. Michelin Tyre plc* 2001 S.L.T. 67 (Scot.).
[101] reg. 5(2).

(a) inspection, testing, adjustment, cleaning and lubrication at suitable intervals;

(b) remedying dangerous defects and preventing access until the defect is corrected;

(c) maintenance and repair to be carried out by trained and competent people; and

(d) the keeping of proper records.

The regulations also require effective and suitable ventilation,[102] reasonable temperatures in working hours[103] and suitable and sufficient lighting.[104] All workplaces, furniture and fittings must be kept sufficiently clean and accumulations of waste avoided;[105] workrooms must be of sufficient size;[106] workstations (including external workstations) and seating must be suitable;[107] and floors and traffic routes must be properly constructed, drained, maintained, kept free of obstructions and materials likely to cause slips, trips or falls. Suitable handrails and guards must be provided in appropriate circumstances.[108] Steps have to be taken so far as is reasonably practicable to prevent injury from falling objects, people falling a distance likely to cause personal injury, and steps must be taken to prevent people falling into tanks, pits or structures containing dangerous materials.[109] Other provisions relate to the suitability of windows, doors, skylights, etc.;[110] the organisation of traffic routes to ensure safety[111] and the suitability of doors and gates,[112] escalators,[113] sanitary conveniences,[114] washing facilities,[115] drinking water,[116]

---

[102] reg. 6.
[103] reg. 7.
[104] reg. 8.
[105] reg. 9.
[106] reg. 10.
[107] reg. 11.
[108] reg. 12.
[109] reg. 13.
[110] regs. 14-16.
[111] reg. 17.
[112] reg. 18.
[113] reg. 19.
[114] reg. 20.
[115] reg. 21.
[116] reg. 22.

clothing accommodation,[117] rest and meals facilities;[118] and the organisation of facilities for disabled people where they are employed.[119]

An Approved Code of Practice accompanies the regulations. There is no specific risk assessment requirement contained in the regulations but the need for one in order to comply with the regulations is fairly obvious.

DUTIES UNDER THE CONTROL OF MAJOR ACCIDENT HAZARDS REGULATIONS 1999[120]

The regulations require every operator to take all measures necessary to prevent major accidents and limit their consequences to people and the environment.[121] The regulations are enforced by the Health and Safety Executive and apply to establishments where specified dangerous substances exist, amongst other things. The regulations require the preparation of a major accident policy,[122] the preparation of safety reports,[123] and the preparation of on-site emergency plans[124] and off-site emergency plans.[125]

The off-site plan has to be prepared by the local authority in whose area the establishment is located and the operator of the establishment has to provide sufficient information to enable the plan to be prepared. Schedule 5 sets out the objectives of on-site and off-site plans, together with the information to be included. It also states the type of information to be provided to the public in the area likely to be affected by a major accident.

It is also possible that certain major accident hazards could fall into the Health and Safety Executive's definition of major incident.[126] The major incident response procedures set out by the Executive[127]

---

[117] regs. 23, 24.
[118] reg. 25.
[119] reg. 25A.
[120] S.I. 1999 No. 743 as amended.
[121] reg. 4.
[122] reg. 5.
[123] reg. 7.
[124] reg. 9.
[125] reg. 10.
[126] See HELA LAC 20/2, March 2000.
[127] *ibid.*

will no doubt be reflected in the emergency plans following consultation with the Executive.

LOCAL AUTHORITY ENFORCEMENT DUTIES

The role of the enforcement authorities is dictated by section 18 of the Health and Safety at Work etc. Act 1974. It is the duty of the Health and Safety Executive to make adequate arrangements for the enforcement of the relevant statutory provisions, except to the extent that some other authority or class of authorities is by any of those provisions or by regulations[128] made responsible for their enforcement.

Section 18(2) states that the Secretary of State may by regulations:[129]

"(a) make local authorities responsible for the enforcement of the relevant statutory provisions to such extent as may be prescribed;

(b) make provision for enabling responsibility for enforcing any of the relevant statutory provisions to be, to such extent as may be determined under the regulations –

(i) transferred from the Executive to local authorities or from local authorities to the Executive; or

(ii) assigned to the Executive or to local authorities for the purpose of removing any uncertainty as to what are, by virtue of this subsection, their respective responsibilities for the enforcement of those provisions;

and any regulations … above shall include provision for securing that any transfer or assignment … is brought to the notice of persons affected by it."

In the past, transfers of responsibility have routinely been made in many cases to reflect the experience of the appropriate enforcing authority. However, the Health and Safety Executive and the

---

[128] Made under s.18(2).
[129] Health and Safety (Enforcing Authority) Regulations 1998, S.I. 1998 No. 494.

Commission have taken serious issue over the general performance of local authorities in enforcing the Health and Safety at Work etc. Act 1974. In its strategy for 2004, the Executive raises the role of local authorities as one of the emerging issues and, at the Chartered Institute of Environmental Health annual Lancaster symposium in July 2003, the chair of the Health and Safety Commission, when comparing the number of local authority health and safety inspections with food inspections, said "I think the status quo is not tenable", stressing however that "it is not about the number of inspections, but about delivering real output and improving health and safety." At the same symposium the head of the Local Authority Unit at the Health and Safety Executive expressed similar concerns. They were no doubt responding to the previous set of performance statistics[130] which showed a continuing overall decline in local authority health and safety performance. A review of the allocation of health and safety duties is on hold at the time of writing. If, as a consequence of this review, it were decided to transfer the whole or part of the current local authority responsibilities to the Executive, then that transfer would take place under the powers in section 18(2). However, the Commission and Executive recognise the valuable role of local authorities in health and safety enforcement and it is likely that improved regional partnerships will be established in an attempt to harness better the skills of the Executive and local authorities.

Section 18(4) of the Act requires every local authority:

(a) to make *adequate* arrangements for the enforcement within their area of the relevant statutory provisions to the extent that they are by any of those provisions or by regulations ... made responsible for their enforcement; and

(b) to perform the duty imposed on them ... and any other functions conferred on them by any of the relevant statutory provisions *in accordance with such guidance as the Commission may give them.*

So, although the Act does not specify what are *adequate* arrangements, the fact that authorities are required to perform their

---

[130] HELA 2002 report, HSE.

duties in accordance with guidance from the Commission makes any such guidance *mandatory*, and it is clear that the Commission is not happy about the performance of many authorities.

Section 18 guidance

This mandatory guidance has been published by the Commission.[131] It contains six key features which are dealt with more fully in the appropriate parts of this book. Those features are:

1.   Enforcement policy and procedures.

2.   Prioritised planning.

3.   Requirement to produce a service plan, including investigation of accidents, complaints, etc.

4.   Requirement to undergo audit and develop an action plan.

5.   Provision of a trained and competent inspectorate.

6.   Requirements in respect of Lead Authority Partnership Schemes (LAPS).

In the view of the Health and Safety Commission, the following elements are essential for a local authority to discharge adequately its duty as an enforcing authority:

(a)   a clear published statement of enforcement policy and practice;

(b)   a system for prioritised planned inspection activity according to hazard and risk, and consistent with any advice given by the Health and Safety Executive and Local Authority Enforcement Liaison Committee (HELA);

(c)   a service plan detailing the local authority's priorities and its aims and objectives for the enforcement of health and safety;

(d)   the capacity to investigate workplace accidents and to respond to complaints by employees and others against allegations of health and safety failures;

---

[131]   "Health and safety in local authority enforced sectors. Section 18", HSC Guidance to Local Authorities, November 2002, HSE.

(e) arrangements for benchmarking performance with peer local authorities;

(f) provision of a trained and competent inspectorate; and

(g) arrangements for liaison and co-operation in respect of the Lead Authority Partnership Scheme.

The section 18 guidance note outlines these elements but recognises that individual local authorities vary in many respects and the guidance is therefore not intended to be so detailed or prescriptive that it requires all local authorities to operate identically. It recognises that individual authorities' priorities are affected by local issues, and their duty to respond to the needs of their own communities and deliver a range of services. Nevertheless, this has not stopped the Commission expressing its concern over the performance of many authorities and undertaking a review of the local authority role in health and safety enforcement.

An inter-authority audit protocol has been issued by HELA which enables the Commission and local authorities to review and monitor the performance of individual authorities.[132]

HELA guidance
The Commission guidance issued under section 18 is supplemented by guidance from HELA in the form of Local Authority Circulars (LACs) and other documents, e.g. on the management of enforcement. Local authorities should consider this guidance in deciding how to comply with their duties under the Act.

Provision of information on enforcement activity, etc.
Local authorities will need to ensure that they devote sufficient resources to the health and safety enforcement function to comply with their duties under section 18, and the Commission will take a view on that issue using information supplied to it by authorities, by consulting HELA and by reviewing reports from inter-authority auditing.

---

[132] "Auditing framework for local authorities' management of health and safety enforcement", September 2002, HSC.

In authorities struggling to meet ever-increasing demands for more and better services, this may present health and safety professionals with a headache as political priorities may well focus on other services. There are few votes in health and safety, and most authorities will give priority to key services such as housing, education and social services. The Commission seems to think that it is the job of health and safety managers to convince politicians of the need to give health and safety a high priority. Whilst this may be fair comment, it ignores the reality of local politics. The really hard bargaining over the resources required for local authority services usually starts about three-quarters of the way through the financial year. Accordingly, if local councillors are to be persuaded to fund health and safety adequately against a raft of competing demands, it is best to try and get their commitment earlier in the financial year. Some ways of influencing their thinking might include the following:

- Reporting cases where authority to prosecute is required and advising elected members of the outcome of prosecutions.

- Reporting on new health and safety legislation and its implications for the local authority as employers and enforcers; any government cost/benefit analysis may help to persuade members of the value of such legislation in protecting employees.

- Advising on the cost of work related accidents and ill-health can provide a graphic indication of the financial and working time losses of inadequate control of health and safety in the workplace.

- Reporting the outcome of audit reviews will inform members of any shortcomings in their authority's performance. If officers can then secure a commitment to improvement, this will help at budget time.

- Any well managed department will review annually its work programme, identifying its statutory obligations and discretionary powers and the resources necessary to undertake the required levels of service. Involving appropriate members in that process, whilst not guaranteeing the ultimate provision of the required resources, will at least ensure an understanding of the service needs and the implications of not providing adequate resources.

- If it can be demonstrated that the available resources are being effectively used and within budget limitations, it is more likely that additional resources may be made available than if it is obvious that a service is poorly managed.

These points may be self evident to most managers but are often worth repeating, even though they offer no guarantees.

Default powers
If a local authority fails to meet its legal obligations under section 18 of the Act, the Secretary of State may, after considering a report from the Health and Safety Commission, cause a local inquiry to be held. If the Secretary of State is satisfied that an authority has failed to perform any of its enforcement functions, he may make an order declaring the authority to be in default. The order may direct the authority to perform its enforcement functions in a specified manner within a specified period of time.[133]

If the authority fails to comply with an order under section 45, the Secretary of State may enforce it or make an order transferring the enforcement functions of the authority to the Health and Safety Executive. In such cases the related expenses are paid by the defaulting authority.

There are no cases to date of the default powers being used, and it is unclear what would happen if the Health and Safety Executive failed to carry out any duties transferred to it. Indeed, it is unclear whether there is any mandatory guidance which the Executive is also obliged to follow, which, if not adhered to, could result in any kind of default action. Is this a case of do what I say, not what I do?

Conflict of interest
The guidance note points out that local authorities can be both duty holders responsible for complying with their health and safety

---

[133] s.45.

obligations as an employer, and enforcing authorities. It reminds authorities that they must be careful to ensure there is no conflict of interest between these respective roles. The HSC believes that clear statements of responsibility and transparency will minimise any potential difficulties. The author is not aware of any local authorities facing problems in this respect. This may be because their responsibilities as employers are usually delegated to separate departments and committees; there are well established consultation arrangements through safety representatives and safety committees; and trade union local links with elected members make it unlikely that any serious transgression of health and safety requirements by an authority would escape the notice of members.

HELA has issued further guidance on the issue of local authority enforcement in premises in which they may have an interest.[134]

ENFORCEMENT POLICY AND PROCEDURES

Section 18 guidance

The Health and Safety Commission states in its section 18 guidance note that it expects local authorities as enforcing authorities to ensure that their approach to enforcement is consistent with the current HSC Statement on Enforcement Policy.[135]

It also says that Ministers wish to ensure that the law protecting the health and safety of workers and the public from risks arising from work activities is effective and that there is a strict and fair enforcement regime. The Statement on Enforcement Policy is in line with the Government's *Enforcement Concordat,* which emphasises better regulation and sets out the principles of good enforcement practice. Those principles,[136] which apply to all enforcement not just health and safety, set out what the government believes businesses and others being regulated can expect from enforcement officers. The principles and procedures of the Concordat can be summarised as set out below.

---

[134] HELA LAC 22/10, April 2000.
[135] Annex 1 of the guidance note.
[136] *Enforcement Concordat,* Cabinet Office, March 1998. Agreed between the Cabinet, Home and Scottish (now the Scottish Executive) Offices and the local authority associations.

Policy
Standards
In consultation with business and other relevant interested parties, clear standards will be drawn up setting out the level of service and performance the public and business people can expect to receive. The standards will be made publicly available, as will the performance measured against them.

Openness
Information and advice on the rules that are applied will be published in plain language and disseminated widely. There will be widespread consultation with relevant bodies.

Helpfulness
As prevention is better than cure, there will be active working with businesses, providing advice and assistance, with contact numbers for people needing information. Applications for statutory approvals will be dealt with promptly and efficiently.

Complaints about service
Accessible, well publicised, effective and timely complaints procedures will be provided for business, the public, employees and consumer groups.

Proportionality
The costs of compliance to business will be minimised by ensuring that action taken is proportionate to the risks. As far as the law allows, account will be taken of the circumstances of individual cases and the attitude of the duty holder when considering action.

Consistency
Enforcement duties will be carried out in a fair, equitable and consistent manner. While inspectors are expected to exercise judgement in individual cases, there will be arrangements to promote consistency, including liaison arrangements between enforcing bodies.

Procedures
Advice from officers will be clear and simple and put in writing, on request, explaining the action required and distinguishing between legal requirements and best practice.

Before any legal action is taken, the opportunity will be given to discuss the circumstances of a case and to resolve any differences, unless immediate action is required. Where immediate action is required, an explanation will be given at the time and confirmed in writing within five to ten working days. Advice will be given on any rights of appeal.

HEALTH AND SAFETY COMMISSION STATEMENT ON ENFORCEMENT POLICY[137]

This enforcement policy statement sets out the general principles and approach which the HSC expects the health and safety enforcing authorities (mainly the HSE and local authorities) to follow. All local authority and HSE staff who take enforcement decisions are required to follow the statement. This appears to be the only element of the section 18 guidance which specifically applies to both local authorities and the HSE.

In allocating resources, enforcing authorities are expected to have regard to the principles set out in the policy, the objectives published in the HSC's and the HSE/Local Authority Enforcement Liaison Committee's (HELA) strategic plans, and the need to maintain a balance between enforcement and other activities including inspection.

The key points of the policy statement can be summarised as follows.

The purpose and method of enforcement

It is the task of enforcing authorities to ensure that duty holders, i.e. employers, employees, the self-employed and others, manage and control risks effectively, thus preventing harm.

The purpose of enforcement is to:

1. Ensure that duty holders take action immediately to deal with serious risks.

2. Promote and achieve sustained compliance with the law.

---

[137] Annex 1 to the section 18 guidance note.

3. Ensure that those who fail in their responsibilities have appropriate action taken against them.

The policy draws attention to various ways of ensuring compliance, e.g. offering information and advice; issuing warnings; serving improvement and prohibition notices; withdrawing approvals; varying licence conditions or exemptions; issuing formal cautions (England and Wales only); and prosecution (or report to the Procurator Fiscal with a view to prosecution in Scotland).

Appropriate investigations are essential before taking enforcement action. In deciding on the resources required, enforcing authorities are expected to have regard to the enforcement principles contained in the policy statement and the objectives published in HSC and HELA strategic plans. This should include striking a balance between investigations and preventative activity.

In referring to the use of Approved Codes of Practice (ACOPs) the policy points out that if anyone is prosecuted for a health and safety offence who did not follow the relevant provisions of an ACOP, the onus is on them to show that they complied with the law in another way. Following advice contained in other guidance documents which is not compulsory, e.g. as issued by the HSC, HSE or HELA, will usually be sufficient to comply with the law. Inspectors will themselves have to consider such guidance in appropriate cases.

In any event, the HSC expects enforcing authorities to use discretion in implementing the law. They have to set out their decision-making process in writing and make it publicly available. Judgements are expected to be made in accordance with the *Enforcement Concordat.*

The principles of enforcement
The HSC believes in firm but fair enforcement, adopting the following principles:

Proportionality
This means relating enforcement action to the extent of a duty holder's failure to comply with the law and the consequent risks to

people arising from that failure. Enforcing authorities are expected
to apply the principle of proportionality in relation to those health
and safety duties which are specific and absolute, and to those
requiring action so far as is reasonably practicable. Deciding what
is reasonably practicable to control risks is a matter of judgement,
and enforcing authorities must take account of the degree of risk on
the one hand and, on the other, the cost in terms of time, money or
trouble needed to avert the risk.[138]

Targeting
This means targeting enforcement on those whose activities give
rise to the most serious risks or where the hazards are least well
controlled. Action should be focused on the duty holders responsible
for the risk. Clear examples of this approach are contained in the
Health and Safety Commission's Business Plan 2002/03 and its
Strategic Plan 2001/04 which identify priority programmes to
tackle the most significant hazards or industries where large
numbers of people are employed and the incidence rate of injuries
or ill health is high. The priorities which will be of particular
concern for local authorities include:

• Falls from height.

• Workplace transport.

• Slips and trips.

• Care homes.

The HSC expects enforcing authorities to have systems for deciding
which inspections, investigations or other regulatory work should
take priority. The duty holder's competence is important as relatively
low risk activities can become dangerous if poorly managed when
compared with higher risk activities which are properly controlled.
However, well managed high hazard activities will require regular
inspection to maintain high standards and meet public expectation
of close control, e.g. large warehouses. The use of a risk-rating
scheme which involves "scoring" premises following inspection

---

[138] See *Martin v. Boulton and Paul (Steel Construction) Ltd.* [1982] I.C.R. 366, DC;
West Bromwich Building Society v. Townsend [1983] I.C.R. 257, applying
Edwards v. National Coal Board [1949] 1 K.B. 704.

according to the identified risks and hazards helps to prioritise the risks and the frequency of inspection required.

Consistency
Consistency does not mean uniformity but taking a similar approach in similar circumstances to achieve similar ends. Duty holders managing similar risks, especially where they have large businesses with multiple outlets, expect a consistent approach to enforcement from the different authorities they may have to work with. In this respect the development of Lead Authority Partnership Schemes (LAPS)[139] can be particularly useful. Achieving consistency is not straightforward. Decisions on enforcement are discretionary and are affected by the degree of risk, the attitude and competence of the employer, any history of incidents or breaches of the law involving the duty holder, previous enforcement history and the seriousness of any breach. Enforcing authorities are expected to have arrangements for promoting consistency in the exercise of discretion, including effective liaison arrangements with other authorities.

As an aid to achieving and maintaining consistency, the guidance issued by the HSC, HSE and HELA should always be considered and followed where relevant. In addition, the use within local authorities of regular standardisation exercises helps to promote consistency between officers.

Transparency
This means helping duty holders to understand what is expected of them and what they should expect from the enforcing authorities. It also means making it clear to duty holders what they are legally obliged to do and why. Inspectors should also distinguish between legal requirements and advice on best practice.

Accountability
Enforcing authorities are accountable to the public for their actions and must therefore have policies and standards against which they can be judged, and procedures for dealing with comments and complaints.

---

[139] See Chapter 2, pp.45-49.

LOCAL AUTHORITY ENFORCEMENT POLICY

There are many examples of good local authority enforcement policies[140] and it may be helpful to provide practical advice on what they might usefully contain. The following are taken from some of the policies examined. There are some common features in many policies which seem to represent good practice. Certain matters should be common to any enforcement policy.

Introduction

This part may contain a commitment to certain principles, for example:

- To follow the mandatory advice of the HSC under section 18 of the Health and Safety at Work etc. Act 1974 together with the guidance produced by the HSC, HSE and HELA.

- Some groups of local authorities work together on the production of model policies and advice. In such cases policies would expect to be consistent, with any departures from these policies being justifiable and, if necessary, subject to consultation with the participating authorities.

- Consultation with other authorities or agencies where there is a shared enforcement role.

- Adherence to the principles of equality and protection of human rights.

Legal status of the policy

The date of approval of the policy and the authorising committee should be stated.

Definition of enforcement

Enforcement should include all action taken to enforce the law. This should include formal action including prosecutions as well as routine inspections, investigations and informal advice.

---

[140] e.g. Coventry City Council Environmental Occupational Health and Safety Enforcement Policy, October 2001; Birmingham City Council Environmental and Consumer Services, Public Health Section Enforcement Policy (2003).

Obtaining copies of the policy

To meet the principle of transparency it is important that people know how to obtain copies of the policy. Appropriate website addresses, telephone numbers and departmental addresses should be included and facilities provided for copies in appropriate foreign languages, braille, on tape and in large type.

General principles of enforcement

These may include reference to the application of the government's *Enforcement Concordat*; the application of fairness, independence and objectivity by enforcement officers; avoiding any undue influence in making decisions; and commitment to proceeding against the right person.

Notifying alleged offenders

In general, those subject to enforcement action, together with any witnesses, should be kept advised on progress of any investigations or legal proceedings unless there are compelling reasons to the contrary. Personal confidentiality should be maintained throughout.

Decisions on the level of enforcement action

It is useful to indicate the range of potential enforcement options.

Prosecution
The criteria used in deciding whether to prosecute should be stated, for example:

(a) the gravity of the offence, e.g. deliberate or persistent breaches of the law; persistently ignoring written warnings or formal notices; seriously endangering health, safety or welfare; assaulting an officer in the course of his duties;

(b) where there has been a significant breach of the law resulting in a serious accident, a case of ill health or serious public concern;

(c) regarding individuals, the authority might seek to prosecute individuals where, in the case of company directors or managers an offence was committed with their consent or connivance;

and regarding employees, where they have been previously warned or where the employer has done all that he can to prevent an employee committing an offence. In considering prosecution, a recognition of human rights principles could usefully be stated.

Formal caution
It should be stated that this approach represents a final warning and reference should be made to the Home Office criteria[141] that there must be sufficient evidence to prove the case, *and* the offender must admit the offence, must agree to the caution and the offender must not have committed the offence before.

Refusal or revocation of a licence
In cases where an activity or business requires a licence, it may be appropriate to consider its refusal or revocation where conditions exist which could result in a prosecution.[142] In such cases it is reasonable to use the same criteria as used to decide if a prosecution is appropriate:

• Deliberate or persistent breach of the law.

• Deliberately or persistently ignoring written warnings or improvement or prohibition notices.

• Endangering to a serious degree, the health, safety or welfare of employees and others.

• Obstructing an officer in the course of his duties.

Court injunctions
In some instances, e.g. where offenders are repeatedly found guilty of breaking the law, it may be appropriate to state that such matters will be considered for injunction proceedings.

Seizure
Where legislation provides for the seizure of goods or equipment, it is relevant to state that such powers will be used.

---

[141] Circ. 18/1994.
[142] See *Prosecution* above.

Forfeiture
Where the law allows, it is relevant to seize and/or prosecute in order to prevent dangerous machinery or equipment being used again in a manner which may adversely affect health and safety.

Formal notice (improvement or prohibition notices)
The type of notice served will depend on the nature of the contravention. An immediate prohibition must apply where circumstances demand. In all other cases a reasonable time must be allowed. Notices must contain details of the appeals procedure.

Written warning and/or advice
It can be stated that, for some offences, letters will be sent setting out the nature of the contravention and the steps that should be taken to rectify things. A deadline for completion should be included, together with a statement of the action that will be considered if matters are not rectified.

Verbal warning and/or advice
Minor breaches of the law may be dealt with by verbal advice. Contraventions should be clearly identified. Advice on how matters should be rectified should be given, including a time-scale for completion. It is a good idea to advise on best practice and also to distinguish between best practice and legal requirements.

Revisiting of premises
It is usual to revisit following service of an improvement or prohibition notice and after giving a written or verbal warning. Revisiting as close as possible to the end of the required time-scale for completion of works or compliance with notice requirements is desirable, otherwise delay may negate the value of the original action.

No action
Where there are breaches of the law, it is only in exceptional cases that contraventions may not warrant any action. Such rare cases might include situations where the cost to the offender far outweighs the impact of the contravention on the general public; the cost of enforcement is wholly disproportional to the benefits in terms of health and safety; or the employer has since ceased to operate his business.

Deciding on prosecution or formal caution

In deciding whether prosecution or formal caution is the appropriate course of action, regard should be had to the tests contained in the "Code for Crown Prosecutors"[143] issued by the Crown Prosecution Service. The tests are:

1. The **evidential test** – there must be enough evidence to provide a realistic prospect of conviction.

2. The **public interest test** – this must be considered where there is enough evidence to provide a realistic prospect of conviction. There may be public interest factors which are in favour of, or are against, prosecution.

A formal caution or prosecution should only proceed if a case has passed both tests.[144]

Decisions on the enforcement action required

For relatively minor breaches of the law, the decisions about the most appropriate form of action can be left to the inspector. Those decisions will usually be based on professional judgement and experience, statutory codes and guidance and advice given by the inspector's own authority or by the HSC and HSE.

For serious offences likely to involve prosecution or formal cautions, it is usual, particularly to ensure both consistency and that a case meets the Crown Prosecution Service tests, for decisions to be made using a formal process involving the inspector, relevant service managers, lawyers and, where necessary, the appropriate council service committee.[145] The decision-making process can usefully be contained in the policy statement.

Liaison with other regulators

In appropriate cases enforcement activities should be co-ordinated with other regulatory bodies, including those cases where responsibilities may overlap. These may include the following:

---

[143] Published in 2000.
[144] See Chapter 4 for more details.
[145] See Chapter 5, pp.194-199 for more details.

- Police – in cases where there has been a work-related death.

- Health and Safety Executive – where there may be breaches of the law in premises accommodating activities controlled by both a local authority and the Executive.

- Fire service – where there is a breach of fire safety law.

- Other local authorities – in cases where an employer has multiple outlets around the country and contraventions at one site may be repeated elsewhere. If an enforcement matter affects a wide geographical area beyond one authority boundary, all relevant authorities and organisations should be contacted without delay.

Considering the views of others

In considering the public interest test, the implications for those affected by an offence, and of the decision whether or not to take enforcement action and, if so, what type, should be taken into account where appropriate. In any event, those affected by an offence should be advised of decisions taken.

Protection of human rights

All policies and enforcement decisions must take account of the provisions of the Human Rights Act 1998 which incorporates the European Convention on Human Rights. In particular, the rights protected by the Convention include:

- Article 6 – dealing with the right to a fair trial.

- Article 8 – the right to respect for private and family life, home and correspondence.

- Article 14 – the prohibition of discrimination on any grounds.

Article 6 may become an issue when dealing with the fairness of a decision as to what action to take following a breach of health and safety legislation, e.g. should a prosecution follow, a caution be given, etc. Fairness can be demonstrated by including the criteria for taking alternative courses of action in the enforcement policy document. If this is formally adopted by the controlling local authority committee, it should be seen as fair and properly considered, any change to the policy will have to be with good reason (and no doubt approved by the committee) and a reference to Article 6 will show that the principles of the Act and the Convention have been adopted.

Although there is little case law dealing specifically with health and safety issues, the Convention will have an impact on health and safety law for the following reasons:

1. It is unlawful for a public authority to act in a manner incompatible with a Convention right, and in this respect an act includes a failure to act.

2. Public authorities are obliged to curtail the activities of corporations and others if they violate the rights of individuals.

3. The courts will be obliged to interpret health and safety legislation so that it is consistent with Convention rights, and in respect of any legal action involving health and safety matters they will be obliged to take account of the Convention in arriving at decisions.

It is usual for local authorities to produce guidelines for their staff on human rights issues as they impact on almost every aspect of an authority's activities. Such guidance may also cover related legislation, including the Data Protection Act 1998 and the Regulation of Investigatory Powers Act 2000 and should be referred to in policy statements.

Publicity

To ensure openness and publicity for enforcement action, authorities may want to publish the names of companies and individuals convicted of health and safety offences. The media should routinely be informed of pending prosecutions and convictions as a means of drawing attention to the need to comply with health and safety legislation and as a deterrent to those likely to disregard their health and safety obligations.

Enforcement policy reviews

All policies must be reviewed at regular intervals to ensure that they remain valid and reflect government guidance, changes in an authority's position and changes in the law. Reviews every one or two years will usually be acceptable.

Enforcement – local authority decision making

Most local authorities are likely to include their decision-making process in their policy statement and an example of the type of contents is given below. They may also wish to make reference to the process used by the Health and Safety Executive, a summary of which is also provided.

The following points are taken from a compilation of local authority processes:

Less serious infringements
Where the infringements suggest that advice, verbal or written warnings, or the service of an improvement or prohibition notice is the appropriate course of action, the investigating officer will decide which option is preferable based on his own judgement, Approved Codes of Practice or other relevant information.

More serious infringements
Where the infringements suggest the need for seizure of equipment or machinery, etc. a formal caution or prosecution, a more rigorous procedure will be required, for example:

- *Seizure* – the inspector and his line supervisor may decide on the appropriateness of this course of action, if necessary securing a warrant to enter premises from a magistrate and securing the attendance of police officers to prevent a breach of the peace.

- *Formal caution or prosecution* – decisions of this kind are likely to be taken by more senior managers following the submission of a report by the inspector. The manager will need to ensure that the proposed course of action is consistent with the "Code for Crown Prosecutors" and Home Office guidance on the cautioning of offenders. If a formal caution is decided upon, it would be usual for this to be issued by the inspector and his line manager. If the alleged offender refuses to accept a formal caution, then a prosecution may follow.

In the case of a proposed prosecution, a senior manager would normally examine the inspector's report to satisfy himself that the evidence meets the relevant criteria before passing it to the authority's solicitors. The solicitors should consider the report to confirm the existence of a *prima facie* case and the justification for prosecution. Authority to proceed with a prosecution would then be obtained in accordance with the authority's established procedure, usually involving a service committee decision.

Health and Safety Enforcement Management Model

An alternative approach to making enforcement decisions can be found in the HSE's Enforcement Management Model.[146] This is infinitely more complex than many local authority decision-making processes but it provides a comparison that local authorities may wish to consider. It is far too lengthy to summarise effectively but it is intended to provide a framework to help HSE inspectors make enforcement decisions in line with the Health and Safety Commission's Enforcement Policy Statement. It sets out the principles inspectors should apply when deciding what enforcement action to take in response to breaches of health and safety legislation. In particular it:

(a) provides inspectors with a framework for making consistent enforcement decisions;

(b) helps managers monitor the fairness and consistency of inspectors' enforcement decisions in line with the Commission's policy;

(c) assists less experienced inspectors in making enforcement decisions.

HSE inspectors apply the principles of the Management Model in

---

[146] Operational Version 3.0, September 23, 2002. See also HSE Operational Circular OC 130/5, July 2002 for more general guidance on the application of the Model.

all their regulatory actions but only formally apply it and record the outcome in certain circumstances, e.g. following the investigation of fatalities and reviewing decisions. The Model recognises its own limitations because of the complex and unique variables that an inspector may encounter in his work.

The HSE believes that the Model provides both the HSE and local authorities with a framework for making enforcement decisions that meet the principles of the Commission, in particular:

(a) promoting enforcement *consistency* by confirming the parameters, and the relationships between the many variables, in the enforcement decision process;

(b) promoting *proportionality* and *targeting* by confirming the risk-based criteria against which decisions are made;

(c) being a framework for making enforcement decisions *transparent,* and for ensuring that those who make decisions are accountable for them; and

(d) helping experienced inspectors assess their decisions in complex cases, allowing peer review of enforcement action, and being used to help less experienced inspectors in making enforcement decisions.

Complex as it may be, the Model is well worth examining as it provides a great deal of supporting information on matters that should be taken into account in deciding the appropriate course of enforcement action.

ACTIVITIES CONTROLLED BY LOCAL AUTHORITIES

The Health and Safety (Enforcing Authority) Regulations 1998[147]

The regulations allocate to local authorities the responsibility for enforcing the Health and Safety at Work etc. Act 1974 and the relevant statutory provisions, subject to specific exceptions, in all premises where the main activity is listed in Schedule 1 to the regulations.

---

[147] S.I. 1998 No. 494, as amended by S.I 1999 No. 3232.

Regulation 2 contains the interpretation and definition of terms used in the regulations and should be read in conjunction with Local Authority Circular 23/4 (rev 2).[148] The circular is intended to assist local authority enforcement officers in resolving questions which may arise, although interpretation is subject to the decision of the courts in any particular case.

Local authorities to be enforcing authorities in certain cases

Regulation 3(1) specifies that where the main activity carried on in non-domestic premises is specified in Schedule 1, the local authority for the area in which those premises are situated shall be the enforcing authority for them. This is subject to certain provisions in regulation 3 and to regulations 4, 5 and 6. The Health and Safety Executive is responsible for enforcement in any other cases, including the common parts of domestic premises. As soon as the main activity comes within Schedule 1, the local authority becomes the enforcing authority and can prosecute, even though the transfer procedures in regulation 5 have not been implemented.[149]

Regulation 3(2) provides that where non-domestic premises *are occupied by more than one different occupier,* each part shall be regarded as being separate premises and must be separately allocated according to the main activity of that separate occupancy, e.g. separate parts of a building occupied by the same employer will not be separately allocated, but a factory canteen staffed by a different occupier such as a catering contractor will generally be enforced by the local authority, even though the HSE is responsible for the other parts.[150] Where any anomalies arise these can be resolved using local transfer arrangements.[151]

Regulation 3(3) provides that while a vehicle is parked in connection with the sale from it of food, drink or other articles, the vehicle together with its pitch shall be regarded as separate premises

---

[148] Health and Safety (Enforcing Authority) Regulations 1998, LAC 23/4 (rev 2), revised December 2000. See also LAC 23/12 and LAC 23/15.

[149] *Hadley v. Hancox* (1987) 85 L.G.R. 402; (1987) 151 J.P. 227; (1987) 151 L.G. Rev. 68.

[150] Subject to reg. 4(2) exceptions covering local authority premises, police and fire authorities, etc.

[151] reg. 5.

allocated to local authorities. In addition to food vendors, the regulation includes a variety of mobile vendors selling items such as flowers, books, paintings and jewellery.[152]

Regulation 3(4) provides that the local authority will be the enforcing authority for the common parts[153] of non-domestic multi-occupied premises, except that if the Executive is the enforcing authority for all other parts of the premises, the Executive shall be the enforcing authority for the common parts.

In the case of land within an airport, regulation 3(4)(b) gives enforcement responsibility to the Health and Safety Executive in respect of common parts which are outside a building (except for car parks, which are allocated to local authorities) or those "airside" to which only passengers and airport employees have access.

Regulation 3(5) allocates complex sites such as tunnel systems,[154] offshore installations,[155] building construction sites, most educational establishments and hospitals to the Health and Safety Executive. At some universities where student residential accommodation is provided at a clearly separate site with little or no educational activities taking place there, those premises would be allocated to local authorities.[156]

Under regulation 3(6)[157] local authorities cannot enforce activities relating to the operation of railways but they are the enforcing authority for the office activities, as defined in regulation 2, of these premises.

Exceptions to the allocations in regulation 3(1)

Regulation 4 makes the Health and Safety Executive responsible for enforcement of the relevant statutory provisions against a list of organisations[158] including local authorities, the police, the

---

[152] Additional guidance on main activity, peripatetic workers and mobile workplaces is given in HELA LAC 23/6.
[153] Defined in reg. 2(1).
[154] As defined in the Channel Tunnel Act 1987.
[155] As defined in the Offshore Installations and Pipeline Works (Management and Administration) Regulations 1995, S.I. 1995 No. 738.
[156] By virtue of Sch. 1, para. 5.
[157] And Sch. 2, para. 12.
[158] reg. 4(3).

Crown, etc. This means, for example, the Executive will enforce control over local authority refuse collection at a supermarket where the local authority is the enforcing authority but has a contract to collect waste. The Executive will also be the enforcing authority for the common parts of multi-occupied premises for which those bodies have duties under the relevant statutory provisions but which may be occupied by someone else, e.g. the common parts of a local authority owned market. The individual market occupiers will still be subject to local authority enforcement.[159]

Regulation 4(4) to (7) provides for enforcement allocation to be modified by other regulations, and the exceptions that are allowed for to be modified through transfers or assignments.[160] All enforcement of section 6 of the Health and Safety at Work etc. Act 1974 is the responsibility of the Health and Safety Executive.[161] From time to time local authority inspectors will become aware of section 6 enforcement issues, i.e. general duties of manufacturers, etc. as regards articles and substances for use at work, and in those cases they should provide their HSE Enforcement Liaison Officer with relevant information to enable those issues to be followed up.[162] In some cases, the Enforcement Liaison Officer may ask for local authority assistance.

Transfer of enforcement responsibility

Regulation 5(1) provides for the transfer of responsibility for enforcing the relevant statutory provisions in respect of any particular premises, part of premises, or any activity carried on there, from the Health and Safety Executive to the local authority or vice versa.

A transfer can only be made by agreement between the two enforcing authorities, or by the Health and Safety Commission.[163] Where a transfer has been made, the authority to whom it has been made must notify those affected by the transfer, presumably

---

[159] For further advice on enforcement in premises occupied by separate occupiers of the specified bodies, see HELA LAC 23/4 (rev 2), December 2000, paras. 31-36.
[160] Under regs. 5 and 6.
[161] reg. 4(4)(a).
[162] See HELA LAC 23/5, November 2000 for details.
[163] reg. 5(2).

including the employer, relevant occupiers in multi-occupied buildings and any employee representatives. Where the Commission has made the transfer, it must give notice to both enforcing authorities concerned.[164]

Assignment of responsibility in cases of uncertainty

Under regulation 6(1) the responsibility for enforcing any of the relevant statutory provisions in respect of any particular premises, part of premises or any activity can be assigned to the Health and Safety Executive or to the local authority. An assignment *can only be made jointly and only where they agree:* first, that uncertainty about their respective responsibilities exists; and secondly, which authority is the more appropriate enforcing authority.

Enforcing authorities should seek legal advice as to whether uncertainty does exist. If the advice is that there is no uncertainty, then the existing allocation under the regulations should stand.[165] For the purpose of removing any uncertainty, either authority can apply to the Commission for the removal of that uncertainty and to have the Commission assign responsibility to whichever authority it considers more appropriate.[166] Approaching the Commission is seen as a last resort but, pending a decision, the authorities are expected to agree the enforcement arrangements for the interim period.[167]

Main activities allocated to local authorities

These activities are contained in Schedule 1 to the regulations. This is set out below together with some explanation and comment.[168]

Schedule 1 to the Health and Safety (Enforcing Authority) Regulations 1998
> "1. The sale of goods, or the storage of goods for retail or wholesale distribution, except –

---

[164] Additional advice on local transfers is contained in HELA LACs 23/8 and 23/1.

[165] Unless a transfer takes place under reg. 5.

[166] reg. 6(2).

[167] The Commission must inform each party of the submissions made by the other, if reasonable to do so: *R. v. Health and Safety Commission, ex p. Spelthorne Borough Council, The Times,* July 18, 1983.

[168] More information and detail is available in HELA LAC 23/4 (rev 2), December 2000 and HELA LAC 23/15, January 2000.

(a) at container depots where the main activity is the storage of goods in the course of transit to or from dock premises, an airport or a railway;

(b) where the main activity is the sale or storage for wholesale distribution of any substance or preparation dangerous for supply;[169]

(c) where the main activity is the sale or storage of water or sewage or their by-products or natural or town gas;

and for the purposes of this paragraph where the main activity carried on in premises is the sale and fitting of motor car tyres, exhausts, windscreens or sunroofs the main activity shall be deemed to be the sale of goods."

This paragraph includes shops and warehouses, the main purpose of which is the sale of goods retail or wholesale, storage for retail or wholesale distribution, or manufacturers' or producers' warehouses containing completed goods for distribution. Premises used for the storage of partly completed goods for additional processing, raw materials and furniture depositories used purely as a storage service would not fall within this category, although auction houses would. However, provided they are not covered by the exceptions at paragraph 1(a) to (c), some of these may be suitable for transfer to local authorities. The exception for container depots associated with docks, airports and railways is not without some doubt as it is said that there should not be a broad assumption that warehouse premises located close to airports (and presumably the other stated premises) are allocated to the Executive. It is likely that many of the stored goods are due for onward distribution to the retail and wholesale sectors.[170] In the case of doubt, queries should initially be discussed with the local Health and Safety Executive Enforcement Liaison Officer.

Timber merchants, metal stockholders and the like will usually be the subject of local authority enforcement, although where there is complex equipment or a significant amount of on-site processing

---

[169] As defined in the Chemicals (Hazard Information and Packaging for Supply) Regulations 2002, S.I. 2002 No. 1689.

[170] HELA LAC 23/4 (rev 2), December 2000, para. 60.

of materials, there may be some justification for transfers to the Health and Safety Executive.

"2. The display or demonstration of goods at an exhibition for the purposes of offer or advertisement for sale."

This paragraph allocates to local authorities premises such as exhibition halls. Any issues of enforcement regarding the safety of any products displayed would be matters for the Executive.[171]

"3. Office activities."

These would be mainly administrative and associated activities including computer software houses.[172]

"4. Catering services."

Where catering services are the main activity at premises, enforcement is allocated to local authorities.

Where catering is a minor activity conducted by the same occupier, enforcement will be by the authority with overall responsibility, e.g. the HSE at factory canteens where manufacturing is the main activity. Catering activities provided in a separate part of the premises and occupied by a catering contractor will be separate premises allocated to local authorities.[173] Premises where the main activity is in-flight catering will be local authority controlled. Local authorities retain their overall responsibility and expertise for enforcement of food hygiene legislation.

"5. The provision of permanent or temporary residential accommodation including the provision of a site for caravans or campers."

This will include hotels, guest houses and residential care homes, sheltered residential accommodation, as well as caravan and tented camping sites. Premises that are nursing homes[174] or where

---

[171] reg. 4(4).

[172] See reg. 2(1).

[173] Unless they are in premises occupied by the bodies stated in reg. 4(3).

[174] Under the Registered Homes Act 1984 in England and Wales, and the Nursing Homes Registration (Scotland) Act 1938.

the main activity is that of providing nursing care are the responsibility of the Health and Safety Executive. Dual-registered premises provide residential care for some occupants and nursing care for others so it is important to ensure the correct allocation of enforcement responsibility.

As distinct from a residential home, a nursing home provides qualified nursing/medical care and the main activity should depend on the main purpose for which the premises are used. HELA LAC 23/2 (rev 2) provides detailed guidance on the factors to be considered in making the distinction. It is vitally important to agree the correct allocation in any doubtful cases because of the vulnerability of the occupants and the relatively higher level of accidents and dangerous occurrences at this type of premises, including slips, trips and falls, burns and scaldings, fires, legionella and the manual handling problems of dealing with elderly and/or infirm residents.

> "6.   Consumer services provided in a shop except dry cleaning or radio and television repairs, and in this paragraph "consumer services" means services of a type ordinarily supplied to persons who receive them otherwise than in the course of a trade, business or other undertaking carried on by them (whether for profit or not)."

Local authorities are responsible for a wide range of activities, including repairs to consumer goods where this is the main activity. The allocation of enforcement responsibility is, however, restricted by the wording of paragraph 6 to:

- Provision of services in a shop in the ordinary meaning of the word[175] so that repairs to certain consumer goods, e.g. shoes, watches, jewellery, clothing and other household goods carried out in shops are the responsibility of local authorities but similar services at a factory unit lie with the Health and Safety Executive.

- Exclude activities where the main purpose is the provision of services to traders and manufacturers, etc., e.g. plant hire or

---

[175]   HELA LAC 23/4 (rev 2), para. 76.

repair, trade-only warehouses or instant print shops. It would be necessary to examine the extent of the trade and non-trade activities in premises providing a dual service to establish the correct enforcement allocation.

Where there is some manufacturing activity supporting a shop, the local authority will be the enforcing authority, unless that activity is provided for a number of shops in which case the main activity will need to be determined to ensure the correct allocation.

"7.   Cleaning (wet or dry) in coin operated units in launderettes and similar premises."

This includes coin operated dry-cleaning units. Laundry services, including service washes carried out by employees, are considered to be consumer services and remain in local authority control.

"8.   The use of a bath, sauna or solarium, massaging, hair transplanting, skin piercing, manicuring or other cosmetic services and therapeutic treatments, except where they are carried out under the supervision or control of a registered medical practitioner, a dentist registered under the Dentists Act 1984, a physiotherapist, an osteopath or a chiropractor."

These activities include a wide range of cosmetic and therapeutic activities including chiropody, acupuncture, ophthalmic opticians and naturopathy. Where one of these practitioners provides the service on his client's premises, enforcement is by the Health and Safety Executive.[176]

"9.   The practice or presentation of the arts, sports, games, entertainment or other cultural or recreational activities except where the main activity is the exhibition of a cave to the public."

This includes sports activities, cinemas and theatres, circuses, riding schools, and dance and drama schools. However, where the

---

[176]   reg. 3, Health and Safety (Enforcing Authority) Regulations 1998, S.I. 1998 No. 494.

main purpose of premises is educational or vocational training similar to that provided in the mainstream educational system, they will remain with the Executive, including their out of hours use for leisure activities.[177] Many of these activities increasingly involve ethnic minority groups and a knowledge of their culture and traditions may be useful in dealing with the control of activities undertaken by them.

Non local authority operated museums, art galleries and theatres are subject to local authority enforcement, although the Commission considers that there may be some of these premises where the risks are such that the Health and Safety Executive should be the enforcing authority. These would have to be the subject of regulation 5 transfers. Valuable advice in respect of local authority enforcement in premises in which they may have an interest is contained in HELA LAC 22/10.

Local authorities are the enforcing authorities for monuments, stately homes and/or grounds owned or operated by the National Trust except where they are essentially working factories or farms. If those are the main activities, local transfers are usually arranged to place them under Health and Safety Executive control.[178]

> "10. The hiring out of pleasure craft[179] for use on inland waters."

"Inland waters" are all waters other than the sea. If the hire company carries out repair and maintenance of its fleet in the out of season period on the same premises, the main activity remains the same.

> "11. The care, treatment, accommodation or exhibition of animals, birds or other creatures, except where the main activity is horse breeding or horse training at a stable or is an agricultural activity or veterinary surgery."

This includes kennels, catteries, animal cosmetic parlours and the

---

[177] HELA LAC 23/4 (rev 2), December 2000.
[178] See HELA LAC 23/9 on the application of the regulations to the National Trust.
[179] Defined in the Docks Regulations 1988 (S.I. 1988 No. 1655), reg. 2(1).

exhibiting of animals in zoos[180] where these are the main activities.

"12. The activities of an undertaker, except where the main activity is embalming or the making of coffins."

Funeral parlours and undertakers' services are allocated to local authorities.

"13. Church worship or religious meetings."

Where the main activity is all forms of religious worship, evangelical and church social events, and where there is church-run accommodation such as hostels, then enforcement is by local authorities.

"14. The provision of car parking facilities within the perimeter of an airport."

"15. The provision of child care, or playgroup or nursery facilities."

This applies to pre-school child care, playgroups or nurseries in non-domestic premises that are not part of a school and are operated independently. Although some pre-school activities involve some education, it is implied that as compulsory education does not begin until the first term after a child reaches five years of age, then the main activity is not educational and the activity is therefore local authority controlled.[181] Although the Health and Safety Executive is normally the enforcing authority in respect of activities on domestic premises,[182] it is recognised[183] that as child care, etc. activities have to be registered with local authorities, then it is sensible for those authorities to be the enforcing authorities.

A detailed explanation of the allocation of activities to the Health and Safety Executive can be found in HELA LAC 23/4 (rev 2).

---

[180] Enforcement in zoos can present particular problems – in the case of *Howletts and Port Lympne Estates Ltd. v. Langridge* (unreported) January 29, 1996 the High Court affirmed the decision of an industrial tribunal to set aside a prohibition notice which sought to ensure the securing of tigers during the cleaning process following the death of a zookeeper.

[181] HELA LAC 23/4 (rev 2), December 2000, para. 89.

[182] reg. 3, Health and Safety (Enforcing Authority) Regulations 1998, S.I. 1998 No. 494.

[183] HELA LAC 23/4 (rev 2).

Appointment of inspectors

Section 19(1) of the Health and Safety at Work etc. Act 1974 states that "every enforcing authority may appoint as inspectors[184]... such persons having suitable qualifications as it thinks necessary for carrying into effect the relevant statutory provisions within its field of responsibility..."

The requirements with regard to appointment of inspectors are contained in Guidance Note 5 of the section 18 guidance issued by the HSC. This requires that local authorities should ensure that they only appoint inspectors who possess the necessary competences to carry out the tasks they will be authorised to do. The HSC considers that those elements of competence are the ones published by the Employers' National Training Organisation and contained in the framework in Annex 2 of the section 18 guidance.

Inspectors not authorised to exercise all of the specified enforcement powers in the Act should either have the competences relevant to the powers for which they have authorisation, or be the subject of adequate and appropriate supervision by inspectors who do have the standards of competence. The section 18 guidance advises that appropriate supervision – *by a colleague who has demonstrated the relevant standards of competence* – may include:

- Being accompanied.

- A follow-up check visit.

- Examination of a sample of letters and enforcement notices.

- "De-briefing" sessions after visits.

- Providing reports on the outcome of some recent visits for review.

In all cases, local authorities and their inspectors are expected to

---

[184] Although the appointment of inspectors is discretionary, the mandatory provisions of the section 18 guidance and the duties contained in the Health and Safety (Enforcing Authority) Regulations 1998, S.I. 1998 No. 494 make enforcement a duty, subject to the power to transfer those duties to the HSE if they are not performed.

have regard to current HELA guidance in carrying out their enforcement functions.

Section 19(2) requires that every appointment of a person as an inspector shall be made by an instrument in writing[185] specifying which of the powers conferred on inspectors are to be exercisable by him. An inspector is only entitled to exercise *such of those powers as are specified* and *only within the field of responsibility of the authority which appointed him.* The powers can be varied by the authority, e.g. when his experience and training justify it, and he must produce his instrument of appointment or an authenticated copy of it when carrying out his duties.

Any inspector seeking to exercise his powers without the necessary authority to do so would be acting *ultra vires* and his actions would be null and void. Indeed, any employer incurring expense on complying with the requirements of an inspector who is not properly authorised may well have grounds to sue both him and his local authority. Having a warrant card is one thing, but to be authorised requires a council minute, duly signed and specifying the purposes for which the inspector is authorised. If an inspector is accompanied by another person,[186] e.g. an expert in the type of workplace or equipment, who is not duly authorised, an employer is within his rights to refuse access to that person, although this is unlikely. However, an inspector should always seek the employer's permission to take with him someone who is not authorised.

Inspectors should *always* carry their authorisation or an authenticated copy of it when exercising their duties. It can be embarrassing to be asked for it[187] and be unable to provide it without going back to the office to collect it. In the meantime, an alleged offender may have disappeared or a contravention that ought to be prosecuted remedied, or vital evidence hidden away. One of the most embarrassing moments can be when an inspector appears in court to give evidence and a sneaky defence lawyer invites the inspector to produce his authorisation to satisfy the

---

[185] "instrument of appointment" is defined in *Campbell v. Wallsend Slipway and Engineering Co. Ltd.* [1978] I.C.R. 1015, DC.
[186] By virtue of s.20(2)(c) of the 1974 Act.
[187] s.19(4).

court of his powers to act – and he cannot produce it! It is suggested that an inspector takes with him the original council minute or certified copy of it, otherwise an embarrassing start to the court case can just make the rest of the proceedings rather difficult to deal with in a calm and considered way.[188] If moving to a different local authority, it is also useful to make sure that the proper authorisation is obtained straight away, using the appropriate officer's emergency powers if a council meeting is some time away.

Powers of inspectors

Section 20(1) of the Health and Safety at Work etc. Act 1974 provides that an inspector may, for the purpose of carrying into effect any of the relevant statutory provisions within the field of responsibility of the enforcing authority which appointed him, exercise certain specified powers,[189] namely:

(a) at any reasonable time (or, in a situation he believes to be dangerous, at any time) to enter any premises which he has reason to believe it is necessary to enter for the purpose of carrying out his duties;

(b) to take with him a constable if he has reasonable cause to believe he will be seriously obstructed in the execution of his duty;

(c) to take with him any other person duly authorised[190] by his authority and any equipment or materials he requires in connection with the power of entry;

(d) to make whatever examination and investigation as may be necessary;

(e) to direct that the premises or any part of them, or anything in them shall be left undisturbed for as long as reasonably necessary to complete the examination or investigation;

---

[188] In the case of *Campbell v. Wallsend Slipway and Engineering Co. Ltd.* [1978] I.C.R. 1015, DC, the Divisional Court held that it was sufficient for the inspector to inform the magistrates' court that he was an inspector and, if required, to produce a certificate (in that case from the Health and Safety Executive) bearing his photograph and stating that he was an inspector authorised to prosecute.

[189] Specified in s.20(2).

[190] For guidance on the authorisation of another person to accompany an inspector, see HELA LAC 22/2, September 2000.

(f) to take any necessary photographs and recordings;

(g) to take samples of any articles or substances in the premises and any samples of the atmosphere in or around the premises;[191]

(h) in the case of any article or substance found in the premises which appears to have caused or to be likely to cause danger to health or safety, he can cause it to be dismantled or subjected to any process or testing but must not destroy or damage it unless this is necessary for the purpose of carrying into effect any of the relevant statutory provisions;

(i) in the case of any article or substance which has caused or is likely to cause danger to health or safety, he can take possession of it and detain it for as long as is necessary to examine it; ensure that it is not tampered with before the examination is completed; ensure that it is available for use as evidence in any proceedings under any of the relevant statutory provisions or any proceedings relating to improvement or prohibition notices;

(j) to require any person whom he has reasonable cause to believe to be able to give any information relevant to any examination or investigation to answer such questions as the inspector thinks fit and to sign a declaration of the truth of his answers;

(k) to require the production of, inspect, and take copies of, or of any entry in, any books or documents which have to be kept by virtue of the relevant statutory provisions and any other books or documents it is necessary for him to see;[192]

(l) to require any person to provide him with facilities and assistance relating to things within that person's control to enable the inspector to exercise his powers;

(m) any other power necessary to enable him to carry into effect any of the relevant statutory provisions.

---

[191] The Secretary of State has power to make regulations concerning the taking of samples, although none have been made to date.

[192] A person cannot be compelled to produce a document of which he would on grounds of professional privilege be entitled to withhold production on an order for discovery in an action in the High Court or an action in the Court of Session.

If an inspector proposes to use his powers in relation to the dismantling, processing or testing of articles or substances, he must, if requested to do so by a person who is present and has responsibilities in relation to the premises, do so in the presence of that person.[193] Before exercising this power, he must consult with appropriate persons to ascertain what dangers, if any, there may be in doing whatever he proposes to do. This could presumably include the owner of the premises or substance, the manufacturers, the Health and Safety Executive and any other experts in relation to the particular matters under consideration. The inspector would in any event have the general duty to take reasonable care of the health and safety of himself and of others under section 7 of the Act.

Where an inspector takes possession of an article or substance found in any premises, he must leave there, either with a responsible person or, if that is impracticable, fixed in a conspicuous position, a notice giving particulars of the article or substance sufficient to identify it. The notice must state that he has taken possession of it under his powers and, if practicable, he must take a sample and give a portion of it, marked so as to identify it, to a responsible person on the premises.[194]

Power to deal with cause of imminent danger

Section 25(1) of the Health and Safety at Work etc. Act 1974 provides that where, in the case of any article or substance found by him in any premises which he has power to enter, an inspector has reasonable cause to believe that, in the circumstances in which he finds it, the article or substance is a cause of *imminent danger of serious personal injury,* he may seize it and cause it to be rendered harmless (whether by destruction or otherwise).

---

[193] Unless he considers that this would be prejudicial to the State – there is nothing to indicate what this means in practice.

[194] s.20(6). Failure by an inspector to give to a responsible person a portion of a sample of dangerous material, taken under subs. (2)(h) and (i), will result in the material taken being inadmissible in evidence even though, if the inspector had taken a sample under the general power in subs. (2)(g), he need not have left a portion of it: *Skinner v. John McGregor (Contractors)* 1977 S.L.T. 83, (Sh.Ct.), but see also *Laws v. Keane* [1982] I.R.L.R. 500, EAT (samples held taken under subs. (2)(g) and no need for inspector to comply with subs. (6)).

Before rendering harmless any article that forms part of a batch of similar articles, or any substance, the inspector must, if practicable, take a sample and give to a *responsible person* at the premises where it was found a portion of the sample marked so as to identify it.[195] As soon as may be after the article has been seized and rendered harmless, the inspector must prepare and sign a written report giving particulars of the circumstances of the action taken, and must give a signed copy to a *responsible person* at the premises and, unless that person is the owner of the article or substance, he must serve a signed copy on the owner. If he cannot, after reasonable enquiry, ascertain the name and address of the owner, he can serve it by giving it to the *responsible person* at the premises.[196]

A customs officer has the power to assist an enforcing authority inspector by seizing and detaining any imported article or substance for 48 hours.[197]

Improvement notices

Section 21 of the Health and Safety at Work etc. Act 1974 provides that if an inspector is of the opinion that a person is contravening one or more of the relevant statutory provisions, or has contravened one or more of those provisions in circumstances that make it likely that the contravention will continue or be repeated, he *may* serve on him an improvement notice. The notice must:

(a)   state that he is of that opinion;

(b)   specify the provisions as to which he is of that opinion;

(c)   give particulars of the reasons why he is of that opinion; and

(d)   require that person to remedy the contravention or the matters causing it within a specified time.

Under section 24, an appeal may be made to an employment (formerly industrial) tribunal. The time period for an appeal must

---

[195]   s.25(2)(a), (b).
[196]   s.25(3).
[197]   s.25A.

be not less than 21 days after service of the notice. This period may be extended by the tribunal. An appeal has the effect of suspending the notice until the appeal has been dealt with or withdrawn.[198]

It is important to prepare the improvement notice properly in accordance with all of the above requirements if it is to be upheld on appeal.[199] Fortunately a health and safety inspector cannot be sued for alleged negligence causing economic damage, in exercising his powers under this section, whether the notice has been issued by the inspector himself, or by an enforcing authority acting on advice given by him.[200] However, this may be small comfort if his employer finds him negligent in the performance of his duties to such an extent as to warrant disciplinary proceedings.

The Health and Safety Executive in its Enforcement Guide[201] gives very useful and practical advice relating to the application of the Health and Safety at Work etc. Act 1974, including advice relating to improvement and prohibition notices.

Deciding to serve an improvement notice
Section 21 of the Act specifies what an improvement notice must contain. This type of notice should only be used where matters can be remedied within the specified time-scale. It cannot be used to require something with no attainable end, e.g. a notice might require the replacement of a broken machinery guard but cannot require that it be maintained in good condition. It can, however, require within a specified time the provision of a maintenance system, as that would be needed to remedy the matters leading to the contravention in the first place.

When notices are served, there should also be discussions with the duty holder on the steps required so that any possible disagreements may be avoided or resolved. This allows the inspector to explain the reasons for his action as well as providing the duty holder with

---

[198] s.24(3)(a).
[199] See *West Bromwich Building Society v. Townsend* [1983] I.C.R. 257 where a notice was quashed because it did not meet the criteria specified in s.21.
[200] *Harris v. Evans* [1998] 3 All ER 522; [1998] 1 WLR 1285, CA.
[201] *Enforcement Guide – Legal Guidance for HSE and LA Inspectors*, September 2001. This is regularly updated and applies to England and Wales.

the opportunity of suggesting alternatives to the notice requirements which may be cheaper or better.

Reference should also be made to any guidance or procedures issued by the enforcing authority, or to the HSE's Enforcement Management Model in deciding the most appropriate course of action.

Prohibition notices

Section 22(1) provides for the service of prohibition notices and "applies to any *activities* which are being or are likely to be carried on by or under the control of any person, being activities to or in relation to which any of the relevant statutory provisions apply or will, if the activities are so carried on, apply."

Subsection (2) provides that if, as regards any of the *activities* to which the section applies, an inspector is of the opinion that, as carried on, or likely to be carried on by or under the control of the person in question, the activities involve or, as the case may be, will involve a *risk of serious personal injury*, the inspector *may* serve on that person a prohibition notice. Although the use of the word *may* allows for discretion, in practice a notice will invariably be required to correct the contravention.

Although "personal injury" is defined[202] as including any disease or impairment of a person's physical or mental condition, what is "serious" is not defined. There does not appear to be any case law on this matter although it is reasonable to presume that the term includes death and major injuries as defined in the RIDDOR Regulations 1995.[203]

Section 22(3) requires that a prohibition notice shall:

(a)   state that the inspector is of the opinion (i.e. a risk of serious personal injury);

(b)   specify the matters which in his opinion give or, as the case may be, will give rise to the said risk;

---

[202] s.53(1) of the Act.
[203] Reporting of Injuries and Dangerous Occurrences Regulations 1995, S.I. 1995 No. 3163.

(c)   where in his opinion any of those matters involves or, as the case may be, will involve a contravention of any of the relevant statutory provisions, state that he is of that opinion, specify the provision or provisions as to which he is of that opinion, and give particulars of the reasons why he is of that opinion; and

(d)   direct that the activities to which the notice relates shall not be carried on by or under the control of the person on whom the notice is served unless the matters specified in the notice and any associated contraventions have been remedied.

A direction contained in a prohibition notice takes effect at the end of the period specified in the notice[204] or, if the notice so declares, immediately.[205] A prohibition notice will have been validly served, even after an accident causing serious personal injury and even though the recipient is in the process of taking steps to remedy the dangerous situation before he resumes the activities.[206] There are a number of practical considerations that may influence the decision-making process:

•   If the activity concerned has not previously been undertaken, the inspector must be of the opinion that it is likely to be undertaken in such a manner that it will involve risk of serious personal injury. If the activity has been carried on but temporarily ceased, he must be of the opinion that it will re-start and state that opinion on the prohibition notice.[207]

•   An opinion that there is a risk of serious personal injury could arise if commonly expected precautions are not in place, e.g. properly qualified personnel, proper training of someone using dangerous machinery.

•   Provided an inspector reasonably believes that there is a risk of serious personal injury, a contravention of a statutory provision is not necessary for a prohibition notice to be valid.

---

[204]   subs. (4)(a). Otherwise known as a deferred prohibition notice.
[205]   subs. (4)(b).
[206]   See *Railtrack plc v. Smallwood, The Times,* February 16, 2001, DC. This related to a railway signal prohibition after an accident.
[207]   *ibid.*

Accordingly, if an inspector is wrong in his opinion that a relevant statutory provision has been breached, the prohibition notice should still be upheld.

- If a direction has been given that an activity must not be continued until specified matters have been undertaken, that direction takes effect immediately or, in the case of a deferred notice, at the end of the specified period.

- A notice should not be deferred if the risk of serious personal injury is so great that immediate action is required to control the risk.

- A deferred prohibition notice may be required in cases where shutting down a process or activity immediately may present other difficulties, e.g. shutting down a process in the middle of an operating cycle. However, in such cases steps must also be taken to minimise risk as far as possible during the shut down.

- Someone in control of a process or machinery who may not be subject to any statutory duty could still be required to close down the activity if there is a risk of serious personal injury. Any failure to do so would be an offence, e.g. a poorly trained employee may carry on or be told to carry on an activity in an unsafe manner without being aware of the risks. In such cases the inspector may need to serve notice on both the employer and employee. The employer or person carrying on the undertaking should be contacted by telephone or in person without delay to advise them of the situation.

During an inspection or investigation, matters giving rise to a risk of serious personal injury should be given preference over other matters. Using a prohibition notice, or seizing and making safe an article or substance, should control the risk to an acceptable level. This should not prevent other appropriate enforcement action from being taken, e.g. prosecution for the contravention.

An appeal may be made against a prohibition notice to an employment tribunal within 21 days, unless the tribunal agrees to extend the time period for appeal because it is not reasonably practicable to appeal within the 21 days.[208]

Registering notices

Improvement and prohibition notices have to be entered in a public register held by the authority,[209] although excluded from that requirement appear to be notices relating solely to the protection of public safety. The details to be included in the register are limited where trade secrets are involved.

Deciding which notice to serve

If there is a risk of serious personal injury, a prohibition notice is the most appropriate course of action, e.g. in the case of unguarded machinery, defective scaffolding or absence of window cleaner's safety harness. There may be cases where both types of notice are needed, e.g. where a prohibition notice deals with the serious risk and an improvement notice requires the taking of steps to prevent the risk recurring.

The inspector must be of the opinion that the preferred form of notice can be justified. Before service it is important to ensure that all relevant evidence has been obtained.

It is important to remember that the service of notices may be followed by prosecution if the recipient fails to comply. Once a notice has been served, the recipient may withdraw any earlier co-operation and further evidence may be difficult to obtain, so evidence such as photographs, statements and samples should be taken prior to service where possible. Any evidence obtained must be sufficient to support the notice on appeal. In the event of an appeal, the inspector must be able to show that he has complied with the statutory requirements for that type of notice; has acted reasonably; has made the wording sufficiently clear; and has properly served the notice.

All the facts to be relied on in the event of a prosecution or an appeal must be carefully noted and preserved, usually in the inspector's notebook.

---

[208] Employment Tribunals (Constitution and Rules of Procedure) Regulations 2001, S.I. 2001 No. 1171 (S.I. 2001 No. 1170 for Scotland).

[209] Environment and Safety Information Act 1988, s.2(3) and the Schedule to the Act.

Provisions supplementary to the service of improvement and prohibition notices
Section 23(2) provides that a notice may (but need not) include directions as to the measures to be taken to remedy any contravention or matter to which the notice relates and any such directions:

(a) may be framed to any extent by reference to any approved code of practice;[210] and

(b) may be framed so as to afford the person on whom the notice is served a choice between the different ways of remedying the contravention or matter.

Notices relating to the structure of a building must not impose more onerous requirements than the building regulations currently in force, and notices which might lead to measures affecting the means of escape in case of fire must not be served unless the inspector has consulted the fire authority.[211] Section 18 of the Fire Precautions Act 1971 requires every fire authority to enforce within their own area the provisions of the Act, and for that purpose to appoint inspectors. The Act also allows[212] fire authorities to arrange with the HSE for some of their functions to be performed by the HSE on their behalf. The Act regulates the issue of fire certificates, provides for the service of improvement and prohibition notices in respect of fire safety, and in many other respects contains similar powers to the Health and Safety at Work etc. Act 1974. It also places a duty[213] upon fire authorities to consult, amongst others, the relevant health and safety enforcing authorities (local authority or HSE) before requiring alterations to buildings. Equally, local authorities are required to consult fire authorities[214] on proposals to erect a building or to make any extension of, or structural alteration to, a building where plans are deposited in accordance with building regulations.

Clearly, although there is no duty to consult on other fire-related

---

[210] s.53(1).
[211] s.53(3), (4). s.53(6) varies the requirements in relation to Scotland.
[212] s.18(2).
[213] s.17.
[214] s.16.

issues, there should be proper consultation arrangements on fire safety issues in premises where a dual legal responsibility applies.

Where an improvement or prohibition notice has been served which is not to take immediate effect, the notice may be withdrawn before the end of the specified time period; and the time period may be extended by an inspector, provided there is no appeal against the notice pending. Accordingly, an immediate prohibition notice, once served, may not be withdrawn. Therefore, if the person on whom it is served resumes the activity before complying with it, or if he complies and then allows the risk of serious injury to recur, he risks prosecution.

Service of notices

Service on an inspector
Section 46(1) of the Health and Safety at Work etc. Act 1974 states that any notice required or authorised by any of the relevant statutory provisions *to be served on or given to an inspector* may be served or given by delivering it to him or by leaving it at, or sending it by post to, his office.

Service on an individual
Sub-section (2) states that any such notice required or authorised *to be served on or given to a person other than an inspector* may be served or given by delivering it to him, or by leaving it at his proper address, or by sending it by post to him at that address.

Service on an individual should be relatively straightforward. It should be addressed to the individual using his *full name* to avoid any claim of incorrect service. The envelope should be similarly addressed.

Service on a limited company or partnership
In the case of a body corporate, the notice may be served on or given to the secretary or clerk of that body.[215] The *name on the notice* should be that of the company, with the registered or principal office being given as its address.[216] Naming the company

---

[215] s.46(3)(a).
[216] s.46(4)(b).

secretary, whether by name or job title, is incorrect as it is the company who has to comply with the notice, not the individual. Addressing the envelope is another matter. The company secretary is the legal face of the company to whom all notices can be sent and who will accept them on behalf of the company. In the case of a partnership, the notice may be served on or given to a partner or a person having the control or management of the partnership, business or, in Scotland, the firm. The proper address in this case is the principal office of the partnership.[217]

Any notice required or authorised to be served on[218] or given to the owner or occupier of any premises may be served or given by sending it by post to him at those premises, or by addressing it by name to the person on or to whom it is to be served or given and delivering it to some *responsible person* who appears to be resident or employed in the premises. Who is a responsible person is not defined but this would presumably involve delivery to the most senior person on the premises at the time of delivery.

A document served by post, i.e. properly addressed, prepaid and posted, will be deemed to have been served at the time which the letter would be delivered in the usual way by post, unless proved otherwise.[219] Such documents should be sent by recorded delivery or special delivery if not delivered by hand so that details of the delivery and a signature can be obtained to confirm proper service. Copies of notices should also be given to employees or their representatives.[220] Deemed service is two days after posting (first class), i.e. two *working* days.

If the service of the document cannot be proved, then any attempt to prosecute for non-compliance with its requirements will fail, as someone cannot be convicted of non-compliance with something that the prosecution cannot show has been served. Similarly, if the actual notice itself is incorrectly addressed, e.g. to the secretary instead of the company, that will give rise to grounds for appeal.

---

[217] s.46(4)(b). s.46(4) allows that the proper address of any person shall be his last known address. This does not apply to companies or partnerships by virtue of s.7, Interpretation Act 1978.

[218] s.46(6). See also s.46(7).

[219] Interpretation Act 1978, s.7.

[220] Health and Safety at Work etc. Act 1974, s.28(8)(b).

Therefore, if time-consuming delays are to be avoided it is best to get notices served correctly first time.

The position of the Crown

Crown bodies must comply with the requirements of the Health and Safety at Work etc. Act 1974 and the relevant statutory provisions. They are, however, excluded from the statutory enforcement provisions, including prosecutions.[221] The Health and Safety Executive does actually issue what are known as Crown improvement and prohibition notices but these are served on the Crown and not on individuals. The notices are non-statutory. Notwithstanding this, Crown *employees* may be prosecuted for health and safety offences[222] and, if convicted, fined or in certain cases imprisoned. This also applies to managers who are personally culpable. In Crown premises only the Health and Safety Executive enforces health and safety legislation.

Appeals against improvement or prohibition notices

Section 24(2) of the Health and Safety at Work etc. Act 1974 provides that a person on whom an improvement or prohibition notice is served may, within such period from the date of its service as may be prescribed, appeal to an employment tribunal; and on such an appeal the tribunal may either cancel or affirm the notice and, if it affirms it, may do so either in its original form or with such modifications as the tribunal may in the circumstances think fit. The appeal period is 21 days[223] from the date of service, although the tribunal may extend that time period on written application.

Where an appeal has been made against a notice, in the case of an improvement notice the bringing of the appeal has the effect of suspending the operation of the notice until the appeal is finally disposed of or, if the appeal is withdrawn, until the withdrawal of the appeal.[224]

---

[221] s.48. For more details see internal circular PIN 45, "Procedures for enforcing health and safety requirements in Crown bodies", Cabinet Office, June 2001.
[222] s.48(2).
[223] Sch. 5, Employment Tribunals (Constitution and Rules of Procedure) Regulations 2001, S.I. 2001 No. 1171 (S.I. 2001 No. 1170 for Scotland).
[224] s.24(3)(a).

In the case of a prohibition notice, the bringing of an appeal has the same effect but only if, on the application of the appellant, the tribunal so directs.

Relevant employment tribunal decisions
Before considering the detailed arrangements for dealing with an appeal, it is useful to consider some relevant decisions made by employment tribunals as these may have a bearing on the preparation of improvement and prohibition notices:

1. The initial onus of proof in an appeal is upon the inspector to satisfy the tribunal that the requirements for an improvement or a prohibition notice are fulfilled. The onus then shifts to the appellant to show, e.g. lack of practicability.[225] Both onuses involve showing the relevant matters *on a balance of probabilities.*[226]

2. In a case which involved the requirement of "conveniently accessible" toilet facilities, an improvement notice was cancelled because, amongst other things, the tribunal held that "something better is always possible but the tribunal had to consider whether the accommodation provided at present contravened the statutory provisions" and it clearly did not.[227]

3. The risks may justify the expensive requirements of an improvement notice. In this case the *company's financial position was irrelevant* to the question whether the tribunal should affirm the improvement notice and evidence as to the company's financial position was not admitted.[228]

4. Where an inspector exercises his discretion under section 23(2) of the Health and Safety at Work etc. Act 1974 to include directions in an improvement notice as to the measures to be

---

[225] See s.40, Health and Safety at Work etc. Act 1974. The language of s.40 clearly imposes a legal burden on the defendant; and interference with the presumption of innocence resulting from this burden is justified, necessary and proportionate: *R. v. Davies* [2002] E.W.C.A. Crim. 2949; [2002] All E.R. (D) 275 (Dec); [2003] J.P.N. 42.

[226] *Readmans Ltd. v. Leeds City Council* [1992] C.O.D. 419.

[227] *A.C. Davis and Sons v. Environmental Health Department of Leeds City Council* [1976] I.R.L. 282.

[228] *Harrison (Newcastle-under-Lyme) v. Ramsey* [1976] I.R.L.R. 135.

taken to remedy any contraventions or matters, he must *confine himself to a requirement which in law he is entitled to impose.*[229]

5.   Whilst it is arguable that, if an improvement notice contains a technical defect in the wording, the notice has no effect and the tribunal should cancel it, this would not be in accordance with common sense or justice where the tribunal is satisfied that there is a plain and continuing breach of the legislation; the tribunal would confirm the notice with the necessary modifications.[230]

6.   A "modification" can include *not only a reduction in the conditions specified but also an extension of them.*[231]

7.   In upholding an immediate prohibition notice dealing with a serious risk of personal injury, the tribunal said that "the evidence of the Engineering Inspector of Factories called as an expert witness ... was to be preferred to the ... appellants' Technical Director..." In this case the tribunal appeared to be more confident about the views of an independent and unbiased witness.[232]

8.   The time period for an improvement notice was extended as it was not practicable for it to be complied with in the time stipulated *because of approvals needed from the local authority* (in this case for a fire escape to be made to its land). The appellants were ordered to pay costs because they had not been diligent in pursuing the necessary approvals from the landowner.[233]

9.   The tribunal held that it was *not possible to rescind the improvement notice and thus order that the requirements of the law should be ignored on the basis of the appellant's good record* as regards maintenance. Nor could the improvement

---

[229]   *Harrison (Newcastle-under-Lyme) v. Ramsey* [1976] I.R.L.R. 135.
[230]   *ibid.*
[231]   *Tesco Stores Ltd. v. Edwards* [1977] I.R.L.R. 120. "Modifications" are defined in s.82(1)(c).
[232]   *Nico Manufacturing Co. v. Hendry* [1975] I.R.L.R.
[233]   *D.J.M. Campion and A.J. Campion v. Hughes* [1975] I.R.L.R. 291.

notice be cancelled on the basis that the matter was a trivial one.[234]

10. The risk may justify the expensive requirements of an improvement notice. In this case the improvement notice was affirmed without modification because the risk of injury could only be reduced to a level which met the statutory provisions by fitting a particularly expensive device. Although the employers' duty under section 2(1) of the Health and Safety at Work etc. Act 1974 is limited to what is *reasonably practicable,* the tribunal considered this was needed to avoid a continuing contravention.[235]

11. In the case of a prohibition notice where there was a risk of serious personal injury but this was not imminent and a time period was specified on the notice, the tribunal decided that, although in extending the time period allowed there remained the risk of serious personal injury, having regard to the *impact of upholding the notice on the company's ability to operate, and its previous good safety record*, it was reasonable to allow a modification and extension of time.[236]

12. An appeal was allowed because the requirements of an improvement notice requiring the appellants to provide protective footwear free of charge, although practicable, was not reasonable because *the expense involved was disproportionate to the risk involved* to the employees merely to ensure that safety shoes were available to wear.[237]

13. If it is alleged that an improvement notice is invalid as being imprecise or vague, the tribunal should not deal with that allegation as a preliminary point without reviewing all the facts. The tribunal should hear the whole case and, if it then concludes that the improvement notice was imprecise or vague, modify the notice to remedy this defect.[238]

---

[234] *South Surbiton Co-operative Society v. Wilcox* [1975] I.R.L.R. 292.

[235] *Belhaven Brewery Co. v. McLean* [1975] I.R.L.R. 370.

[236] *Otterburn Mill v. Bulman* [1975] I.R.L.R. 223.

[237] *Associated Dairies v. Hartley* [1979] I.R.L.R. 17.

[238] *Chrysler (U.K.) Ltd. v. McCarthy* [1978] I.C.R. 939, DC.

14. The tribunal *does not have the power to modify an improvement notice by adding allegations of breach of further statutory provisions*, even though asked to do so by the inspector who served the notice.[239]

15. The decision on appeal of a tribunal has been held *not to be binding on a criminal court* in a prosecution for non-compliance with the improvement or prohibition notice.[240]

There is, of course, one important lesson to learn from decisions of this kind – get the notice right first time!

EMPLOYMENT TRIBUNALS

Notice of appeal

When an improvement or prohibition notice is served, it should be accompanied by information advising the recipient how they may appeal to an employment tribunal. The manner in which appeals are dealt with is contained in the Employment Tribunals (Constitution and Rules of Procedure) Regulations 2001 and, more particularly, Schedule 5 – The Employment Tribunals (Improvement and Prohibition Notices Appeals) Rules of Procedure.[241] The person lodging the appeal must send details of the grounds of appeal to the tribunal within 21 days of service of the notice.[242] The tribunal may extend the time period where it is satisfied on a written application, made before or after the expiration of the 21 days, that it was not practicable for an appeal to be brought

---

[239] *British Airways Board v. Henderson* [1979] I.C.R. 77 (tribunal decision). *Cf. West Bromwich Building Society v. Townsend* [1983] I.C.R. 257; *The Times,* January 3, 1983. s.24(2) of the Health and Safety at Work etc. Act 1974 only allows a tribunal to modify, not add to, a notice.

[240] *Paull v. Weymouth and Portland B.C.* (1979) 25 *The Inspector* 67; (1979) *Environmental Health* 235 (Crown Court). See also s.33(1).

[241] S.I. 2001 No. 1171. The regulations relate only to England and Wales and prescribe the rules of procedure for proceedings before employment tribunals. They define the qualifications of the President of the Employment Tribunals and the panels of members, their composition, and objectives (including equality, saving expense, proportionality and fairness). Schedule 5 contains a code for the initiation, conduct and determination of appeals against improvement and prohibition notices. Similar provisions for Scotland are contained in S.I. 2001 No. 1171, as amended by S.I. 2001 No. 1460.

[242] *ibid.,* Rule 2(1).

within that time.[243]

A copy of the appeal must be sent by the tribunal to the enforcing authority, together with the address to which notices and communications must be sent.[244] If a decision has been made by the enforcing authority to prosecute, or such a decision is very likely, the tribunal should be informed. It is then likely to put the hearing on hold until after the prosecution.

Application for the suspension of a prohibition notice
Where an appeal has been brought against a prohibition notice and a written application is made to the tribunal for a direction suspending the operation of the notice until the appeal is finally disposed of or withdrawn, the application must be sent to the tribunal secretary. It must contain the grounds on which the application is made and a copy must be sent by the tribunal to the enforcing authority.[245]

Where a notice is suspended on appeal and the compliance date has expired before the hearing, the tribunal will be able to insert a new compliance date if it upholds the notice. However, if the appeal is withdrawn after the compliance date has expired, the duty holder may be in immediate breach of the notice. If this situation is likely, the tribunal should be informed so that it can deal with the matter appropriately.

Whilst an improvement notice is the subject of an appeal, the enforcing authority cannot extend the time period. Provided the time period on the notice has not expired, the enforcing authority can withdraw the notice but the appellant may ask the tribunal to

---

[243] S.I. 2001 No. 1171, Rule 2(2). See also *G.W. Fantarrow v. Leworthy,* Newcastle upon Tyne, October 15, 1979, case no. HS/21905/79 folio ref. 9/HS/1/13. The tribunal dismissed an application for an extension of the time period for appealing against an improvement notice where the notice of appeal was dated 40 days after service of the notice. It said that the time limit was a recognition of the fact that improvement and prohibition notices affect the health and safety of persons at work and may involve serious questions of danger to health or personal injury. It rejected arguments that the appellant was negotiating to get the notice withdrawn, that information was being sought on the cost, and the delay was also due to pressure of work.
[244] *ibid.,* Rule 3(a), (b).
[245] *ibid.,* Rule 4.

award costs against the authority. If it is decided to withdraw a notice after the expiry date has passed, the tribunal should be informed that the appeal will not be opposed. As the appellant may ask for costs, it may be appropriate to discuss the matter with the appellant to see if he is prepared to forgo costs if the notice is withdrawn. Alternatively, a proportion of the costs might be agreed by the local authority, especially if it is at fault in serving the notice, e.g. too hastily.

Preparing for the tribunal hearing

The tribunal has powers which help the parties in preparing their cases. These powers are exercised in accordance with the objective to deal with cases justly,[246] including ensuring that the parties are on an equal footing; saving expense; and dealing with cases in ways which are proportionate to their complexity, expeditiously and fairly. Accordingly, tribunals have wide discretion in the conduct of proceedings.

The notice of appeal should set out the details of the specific requirements that are the subject of appeal, together with the grounds of appeal. If those grounds are not sufficiently clear then the appellant can be notified that further particulars of those grounds are required, together with any relevant facts and arguments. If the authority is dissatisfied with the response, it may apply for an order from the tribunal seeking further information.[247]

Power to require attendance of witnesses and production of documents, etc.
A tribunal may on the application of a party to proceedings, either by notice to the Secretary or at the hearing, require a party to provide in writing further particulars; grant disclosure or inspection of documents; and require the attendance of any person as a witness or require the production of any relevant documents.[248] Failure to provide certain documents without reasonable excuse may result in a fine upon summary conviction.[249] Both the enforcing

---

[246] See the Employment Tribunal (Constitution and Rules of Procedure) Regulations 2001, S.I. 2001 No. 1170.
[247] *ibid.,* Sch. 5, Rule 5.
[248] *ibid.,* Sch. 5, Rule 5(1)(a) to (c).
[249] *ibid.,* Sch. 5, Rule 5(4).

authority and an appellant can make an application for an order, so the basis of a decision to serve an improvement or prohibition notice can come under close scrutiny. An order may need to be sought where either party:

(a) needs to know the nature of the case they have to defend;

(b) wants to prepare the best possible case in defence of their position;

(c) wants to avoid being taken by surprise;

(d) wishes to determine the relevant issues between them, which may in turn determine the extent of disclosure required.[250]

The tribunal has the same powers as a county court to require either party to make disclosure.[251] It should also be recognised that standard disclosure includes both documents on which a party relies in support of its case and documents which may adversely affect its own or another party's case. Specific disclosure of documents may be ordered where it is possible to identify specific documents which are necessary to ensure a fair hearing.

Where someone who is not a party to an appeal holds documents which are required, e.g. notices may have been served on joint occupiers or owners of a building, or several contractors most of whom have not appealed, it may be necessary to rely on other disclosure powers.[252]

Witness evidence
It is quite usual to rely on the evidence of the inspector and any relevant witnesses. That evidence should be provided in the form of a witness statement, although the tribunal is not a criminal court and therefore the provisions of the Criminal Justice Act 1967 are not applicable. The inspector's statement in support of his action should give the reasons for his decision. These may include:

•   The circumstances he found that gave rise to the risk(s)

---

[250] *C. White v. University of Manchester* [1976] I.C.R. 419.
[251] Employment Tribunal (Constitution and Rules of Procedure) Regulations 2001 (S.I. 2001 No. 1170), Sch. 5, Rule 5(1)(b).
[252] Under the Civil Procedure Rules 1998, S.I. 1998 No. 3132 as amended.

covered by the requirements of his notice.

• The potential injuries or damage that could occur if the conditions were not rectified.

• The response of the appellant to the circumstances giving rise to the notice, e.g. did he accept the risk, did he indicate a willingness or otherwise to act positively to deal with the situation?

• The number or people exposed to the circumstances giving rise to the notice.

• The reasons for taking the chosen course of action rather than any other options that might have been available.

The inspector should seek to persuade the tribunal that the action taken by him is more appropriate than any of the options that may be available to the tribunal.

Attendance of witnesses
A willing witness is always likely to be a better witness than a reluctant one . Therefore it is always best to try and get any witness an inspector wishes to call to attend voluntarily. However, if that is not possible it will be necessary to ask the tribunal to order the attendance of the witness.[253] An application should be made to the Secretary to the tribunal as early as possible. The tribunal will need to be satisfied that the witness will not attend voluntarily, e.g. the witness has been interviewed, a statement taken but he refuses to attend to give evidence. In the case of an employee, it may be preferable to apply for an order to avoid any risk of discrimination by the employer against the employee.

Many tribunal or court cases have been called off at the last minute due to unavailability of witnesses. It is therefore desirable to ascertain the availability of all witnesses well in advance of any hearing and advise the tribunal accordingly. If possible, the tribunal should be informed of the estimate of the time required to present the authority's case so that it can make similar enquiries of the other party and allocate an appropriate amount of time at a suitable

---

[253] Employment Tribunals (Constitution and Rules of Procedure) Regulations 2001, Sch. 5, Rule 5(1)(c).

date or dates. It would be helpful if the tribunal could be asked not to list a case until the availability of all witnesses is known.

Employment tribunal hearings

Any hearing of an appeal must be heard by a tribunal composed in accordance with the relevant provisions of the Employment Tribunals Act 1966.[254] Subject to certain provisos, the appeals must take place in public. However, a tribunal may sit in private:[255]

(a) for the purpose of hearing evidence from any person which in the opinion of the tribunal is likely to consist of:

    (i) information which he could not disclose without contravening a prohibition imposed by or by virtue of any enactment;

    (ii) any information which has been communicated to him in confidence or which he has otherwise obtained in confidence from another person; or

    (iii) information the disclosure of which would cause substantial injury to any undertaking of his or in which he works; or

(b) if it considers it expedient in the interests of national security.

Written representations

Any party wishing to submit written representations for consideration by a tribunal at the hearing of the appeal must present those representations to the tribunal secretary not less than seven days before the hearing and, at the same time, send a copy of it to the other party.[256] The term "written representations" is not defined but may be taken to include anything contained on the appeal form and any response to it.

A tribunal must consider any written representations before dismissing a case, and must also consider such representations if

---

[254] Employment Tribunals (Constitution and Rules of Procedure) Regulations 2001, Sch. 5, Rule 7.

[255] *ibid.,* Sch. 5, Rule 7(3).

[256] *ibid.,* Sch. 5, Rule 8. The service of any notices, etc. under the rules of procedure has to be in accordance with Rule 15.

either or both of the parties fails to appear. The tribunal will adjourn the hearing if it considers that further inquiries should be made into any non-appearance.

Although witness statements are not considered to be written representations, a tribunal has sufficient discretion to admit them in the absence of the witness. However, in such cases the statement may carry little weight if there is a conflict of evidence.

Procedure at the hearing
At any hearing of or in connection with an appeal, a party can represent himself or be represented by any other person. That may be a lawyer or someone who is not legally qualified. In the case of local authorities there is no clearly defined practice and many cases have involved only the inspector who initiated the notice. It is nevertheless likely that an inspector will have sought his authority's professional legal advice before appearing before a tribunal.

Legal aid is not available to appellants in tribunal proceedings, only for initial advice and help and for any subsequent appeal from a tribunal by an individual, not a company. At the hearing either party is entitled to:[257]

(a) make an opening statement;

(b) give evidence on his own behalf;

(c) call witnesses;

(d) cross-examine any witnesses; and

(e) address the tribunal.

In presenting an *opening statement,* the local authority inspector should bear in mind the following points:

1. The statement should state the principal reasons for serving the notice in question. This may involve repeating the reasons for service and the opinions required to be given on the notice

---

[257] Employment Tribunals (Constitution and Rules of Procedure) Regulations 2001, Sch. 5, Rule 9(1).

at the time of service.[258] It may also need an explanation of any risks associated with contravention of the relevant statutory provisions.

2. It should advise the tribunal of the witnesses, documents, photographs or any other samples or evidence on which he intends to rely. It is of course essential to ensure well before the date of the hearing that all of this information will be available, including checking that witnesses actually intend appearing.

3. As the tribunal deals with a wide variety of employment matters, not just improvement and prohibition notice appeals, it is as well not to assume that it is fully familiar with the relevant law and offer to outline it as it relates to the service of improvement and/or prohibition notices.

4. As there will be no formal transcription of the proceedings, it may be helpful to have someone present from the inspector's department to take detailed notes of the proceedings. This will be helpful in the event that it is decided to appeal against any decision of the tribunal. It can also be helpful in developing training material for inspectors.

In *giving evidence* the inspector can expect to be cross-examined by the appellant, and re-examined by anyone presenting his case, on any matters requiring clarification as a result of the cross-examination. The inspector's witnesses will be subject to the same procedure.

If a party fails to appear or be represented at the hearing, the tribunal may deal with the appeal in the absence of the party or may adjourn the hearing to a later date. Before dealing with the appeal in the absence of a party, the tribunal must consider any written representations submitted by that party.[259] A tribunal may also require any witness to give evidence on oath or affirmation.[260]

In *addressing the tribunal* the person presenting the case for the authority, i.e. the inspector or another person, should summarise

---

[258] As per ss.21 and 22 of the Health and Safety at Work etc. Act 1974.

[259] Sch. 5, Rule 9(2).

[260] Sch. 5, Rule 9(3).

the evidence in support of the authority's decision to serve the notice and seek to deal with any points made against it by the appellant.

Decision of the tribunal

The decision of a tribunal must be recorded in a document signed by the Chairman and must contain the reasons for the decision.[261] A copy must be sent to each party and the document entered in the public register.[262] The tribunal can either cancel or affirm the notice[263] and, if it affirms the notice, it may do so either in its original form or with such modifications as it thinks fit. Any purely technical problem with the notice should result in an amendment, unless the tribunal decides that the inspector's opinion was so unreasonable as to be insupportable.[264]

If the notice has been suspended pending determination of the appeal, the date for completion of works or other steps may have expired. In such a case the tribunal will need to consider modifying the compliance date and the inspector will need to express an opinion about the associated risks in order to assist the tribunal.

As a tribunal has the power to make an order for one party to pay the agreed costs of the other party, or in default of an agreement the amount of costs assessed by way of a detailed assessment,[265] the costs should be calculated in advance of the hearing and given to the appellant at its conclusion.

Challenging a tribunal decision

A tribunal has the power, on the application of any of the parties, to review and revoke or vary by certificate any of its decisions on the grounds that:[266]

(a) the decision was wrongly made as a result of an error on the part of the tribunal staff;

---

[261] Sch. 5, Rule 11(2).
[262] Sch. 5, Rule 11(3). The reasons must be omitted from the register in any case in which evidence has been heard in private and the tribunal so directs.
[263] Health and Safety at Work etc. Act 1974, s.24(2).
[264] *Chrysler (U.K.) Ltd. v. McCarthy* [1978] I.C.R. 939.
[265] Sch. 5, Rule 13.
[266] Sch. 5, Rule 12(1).

(b) a party did not receive notice of the proceedings leading to the decision;

(c) the decision was made in the absence of a party;

(d) new evidence has become available since the making of the decision, provided that its existence could not have been reasonably known of or foreseen; or

(e) the interests of justice require such a review.

An application can be made at the hearing or in writing within 14 days of the date of entry in the public register. An application can be refused by the Chairman of the tribunal presiding over the case if in his opinion it has no reasonable prospect of success. He must state the reasons for his opinion.

If, after a review by the tribunal, a party wishes to challenge the decision, then an appeal, on a point of law, must be made within three months of the decision being supplied to the parties to the High Court.[267] There may be reason to appeal on a point of law if the tribunal decision is perverse. However, if the appeal has been properly conducted and all the evidence properly considered, it may be difficult to demonstrate that the decision is wrong in law. The tests to be applied to an appeal in the High Court will be similar to those considered by the tribunal.

---

[267] Civil Procedure Rules 1998, S.I. 1998 No. 3132 as amended.

Chapter 4

INVESTIGATION AND PROSECUTION OF OFFENCES

Published information on inspection and enforcement by local authorities[1] reveals that, in 2001/02, 325 informations were laid in Great Britain giving rise to 307 convictions, a conviction rate of 94%. Both figures were the lowest in five years. The majority of the convictions were under the Health and Safety at Work etc. Act 1974 and the Management of Health and Safety at Work Regulations 1999.[2] The average level of fine imposed on conviction was £3,134, a fall of 20% from the previous year. The average level of fines is continuing to fall.

OFFENCES UNDER THE HEALTH AND SAFETY AT WORK ETC. ACT 1974

The nature of an investigation into an alleged offence under the Act may vary depending on the kind of offence involved. It may therefore be useful to outline the offences before dealing with the investigation process.

Offences

Section 33(1) states that it is an offence for a person:

(a) to fail to discharge a duty to which he is subject by virtue of sections 2 to 7 (*general duties of employers, employees, the self-employed and manufacturers*);

(b) to contravene section 8 or 9 (*duty not to interfere with or misuse things, duty not to charge employees for certain things*);

(c) to contravene any health and safety regulations or any requirement or prohibition imposed under any such regulations (*e.g. Health and Safety at Work Regulations 1999*);

---

[1]   *HELA Health and Safety Activity Bulletin 2003*, October 2003, HSC.
[2]   S.I. 1999 No. 3242.

(d) to contravene any requirement imposed by or under regulations under section 14 (*power of the Health and Safety Commission to direct investigations and enquiries*) or intentionally to obstruct any person in the exercise of their powers;

(e) to contravene any requirement imposed by an inspector under section 20 or 25 (*improvement notices and power to deal with cause of imminent danger*);

(f) to prevent or attempt to prevent any other person from appearing before an inspector or from answering any question to which an inspector may by virtue of section 20(2) require an answer;

(g) to contravene any requirement or prohibition imposed by an improvement notice or a prohibition notice (including one modified on appeal);

(h) intentionally to obstruct an inspector in the exercise or performance of his duties;

(i) to contravene any requirement imposed by a notice under section 27(1) (obtaining of information by the Commission, Executive, etc.);

(j) to use or disclose any information in contravention of section 27(4) or 28;

(k) to make statements known to be *false or reckless*;

(l) intentionally to make a *false entry* in any register, book, notice or other document required by or under any of the relevant statutory provisions ... or, with intent to deceive, to make use of any false entry;

(m) with *intent to deceive*, to use a document issued ... under any of the relevant statutory provisions ... or make or have in his possession a document which is calculated to deceive, e.g. a forgery;

(n) falsely to pretend to be an inspector;

(o) to fail to comply with a court order requiring offences to be remedied.

A person found guilty of an offence under paragraph (d), (f), (h) or (n) of subsection (1), or of an offence under paragraph (e) consisting of contravening a requirement imposed by an inspector under section 20, is liable on summary conviction to a fine not exceeding level 5, all other cases being capable of being tried "either way".

Offences due to the fault of another person
Section 36(1) provides that, where the commission by any person of an offence under any of the relevant statutory provisions is due to the act or default of some other person, that person shall be guilty of the offence, and a person may be charged with and convicted of the offence by virtue of this subsection whether or not proceedings are taken against the first-mentioned person. So, for example, if an employee removed the guard to an item of machinery, the employer could be in breach of sections 2 and/or 3 of the Health and Safety at Work etc. Act 1974. The person responsible for removing the guard could be prosecuted under section 36 but that would not stop the employer being prosecuted as well if the particular circumstances revealed that the employer was also at fault in not identifying and rectifying the situation.

Offences by bodies corporate[3]
Section 37(1) provides that, where an offence under any of the relevant statutory provisions committed by a body corporate is proved to have been committed with the consent or connivance of, or to have been attributable to any neglect on the part of, any director, manager, secretary or other similar officer of the body corporate or a person who was purporting to act in any such capacity, he as well as the body corporate shall be guilty of that offence and shall be liable to be proceeded against and punished accordingly.

---

[3] There is not a great deal of relevant case law but for interesting discussions on the issue of corporate manslaughter, see "Corporate crime in the health and safety field", *Encyclopedia of Health and Safety at Work – Law and Practice*, Issue 6, December 1999; "A case for culpability – is the law failing us!" Barrett, Julie, *Environmental Health Journal*, November 2003; and *Corporations and Criminal Responsibility*, Wells, C., Oxford University Press, 2002.

This section allows action to be taken against the most senior managers, i.e. those with the primary responsibility for running the organisation, where the corporate body is in breach of its statutory duties. This section, unlike section 36, is particularly relevant for dealing with situations where the most senior managers have failed to establish safe systems of work or allowed safety to deteriorate through their neglect or connivance. In large organisations it can be difficult to determine exactly where the fault lies, but in small organisations with a smaller and less complex management structure it may be easier to establish which manager is at fault.

The position concerning Approved Codes of Practice
Under section 17(1) of the Health and Safety at Work etc. Act 1974, failure to observe any provision of an Approved Code of Practice shall not of itself render a person liable to any civil or criminal proceedings. However, where in criminal proceedings a party is alleged to have committed an offence by contravening a provision for which there was an Approved Code of Practice at the time of the allegation, then any provision of the Approved Code of Practice which a court considers to be relevant shall be admissible in evidence. Approved codes have a status similar to the Highway Code. HSE Guidance Notes may also indirectly give rise to a duty of care and, if an employee does not warn his employees of the dangers referred to in the notes, he may be liable in negligence.[4]

INVESTIGATIONS

There are several types of investigation which may be conducted, each one depending on whether an offence has been, or is likely to be, committed:

1.   If an *offence has been committed*, e.g. an untrained employee is found operating dangerous machinery or an employee is not wearing essential protective clothing provided for him. Although the inspector in such cases should have enough information to take the appropriate type of action to deal with

---

[4]   *Burgess v. Thorn Consumer Electronics (Newhaven) Ltd.*, *The Times*, May 1, 1983.

the breach, he must, in securing the evidence, ensure that he has acted within his statutory powers and recorded the evidence on which he will rely in any further proceedings.

2. If an offence *may have been committed*, e.g. a complaint has been made that an employer is not training his employees prior to the use of dangerous machinery or is refusing to provide protective clothing in situations where it is clearly needed. In this type of case, the inspector will be trying to form an opinion on whether an offence has occurred and he may need to examine records, interview people and take copies of records and other documents.

3. An inspector decides that the safety record of an employer is so poor that he needs to conduct observations to determine whether the employer is *likely to breach his obligations*. In such cases, the inspector must conduct himself in a way that he is not in breach of his investigatory powers[5] whilst still obtaining and recording all relevant information.

There are a number of possible enforcement outcomes to an investigation:

1. A decision that there is no case for any action.

2. Any breaches found should be dealt with by way of an improvement or prohibition notice.

3. A "formal caution" should be issued.

4. A prosecution is the most appropriate course of action.

The precise form of action will be determined by reference to the authority's enforcement policy.[6] Collecting evidence and conducting investigations in the correct manner are essential to ensure that justice is served and the risk of problems and legal challenges at a later stage is minimised.

---

[5] e.g. Human Rights Act 1998, Regulation of Investigatory Powers Act 2000.
[6] See Chapter 3, pp.96-101.

The impact of the Criminal Procedure and Investigations Act 1996

Investigation rules

The Act contains rules that must be followed by those undertaking criminal investigations. Accordingly, it applies to investigations conducted for the purpose of the Health and Safety at Work etc. Act 1974. It requires proper case management and administration procedures, including the compulsory disclosure of information to ensure fair trials. The Home Office has also issued a related Code of Practice containing mandatory guidance which investigators must take into account.[7] The Act places three main duties on investigators:

1. The identification of officers involved in the investigations.

2. The retention of the evidence.

3. The disclosure of the evidence.

Anyone, not just a police officer, who is charged with the duty of conducting an investigation with a view to ascertaining whether a person should be charged or whether a person is guilty must have regard to any relevant provisions of the Code of Practice.[8] The provisions of the Act become operative as soon as a criminal investigation starts. It has been established that an investigation may have begun prior to an offence having been committed.[9] The provisions are therefore relevant when surveillance has commenced in respect of offences, some of which may have occurred before the start of the surveillance and some afterwards.

A court will determine on the basis of the facts in each case when an investigation may have begun, but it is important for an

---

[7] Criminal Procedure and Investigations Act 1996, Code of Practice, brought into force by the Criminal Procedure and Investigations Act (Code of Practice) (No. 2) Order 1997, S.I. 1997 No. 1033. The relevant Codes of Practice issued in accordance with the Police and Criminal Evidence Act 1984 are also intended to ensure fairness in criminal investigations by ensuring that the evidence obtained is admissible. The codes relevant to local authority investigations deal mainly with the searching of premises, and treatment and questioning of suspects and witnesses.

[8] *ibid.*, s.26.

[9] *Uxbridge Magistrates' Court, ex p. Patel (DC), The Times*, December 7, 1999.

inspector to make his own decision on this as the date will affect the extent to which the prosecution may be required to produce and disclose material obtained during the investigation, for example:

(a) the date of any initial observations prior to the service of an improvement or prohibition notice;

(b) the date of any complaint or receipt of other information leading to those observations;

(c) the date of service of an improvement or prohibition notice;

(d) the date of any alleged breach of the notice.

The Act specifies different roles for those involved in criminal investigations and these must be reflected in the administrative and practical guidance which enforcing authorities should establish as part of their enforcement procedures. Those roles are as follows:

1. *The investigator.* This will include any officer involved in conducting the investigation. He must record and retain information and material in accordance with the Code of Practice.

2. *The officer in charge.* This will usually be the immediate line manager of the inspector. He will be responsible for ensuring that the investigation is carried out in accordance with the law, the authority's policies and the administrative and operational procedures that exist. He *must retain overall responsibility* for ensuring things are done correctly, even though he may delegate specific tasks.

3. *The disclosure officer.* The officer has responsibility for examining and revealing to the prosecutor any material that has been created or recorded as a consequence of the investigation. He must ensure that the rules relating to primary and secondary disclosure are adhered to and that prosecution material is subject to ongoing review and disclosure.[10] He is obliged by the Code of Practice to produce relevant material to the prosecutor and also certify when this has been done.

---

[10] ss.3-9.

4. *The prosecutor.* This will be a local authority lawyer or legal executive with responsibility for conducting the proceedings and who will have been authorised in this respect.

Most of the above roles are easy to allocate to individuals, the one exception being the disclosure officer. The officer must be objective and independent of the investigatory role and therefore should not be the investigating officer. The prosecutor may often be the preferred choice for disclosure officer as he will be presenting the case and is arguably best placed to see the loopholes and pitfalls in the evidence presented to him.

In some cases the investigating and prosecuting officer are one and the same. Whatever the arrangements, it important to ensure that a well documented procedure exists which meets the requirements of the Act and its Code of Practice and will withstand any scrutiny in a court of law.

Those involved in a criminal investigation must have regard to specific requirements of the Code of Practice, particularly:

1. All criminal investigations should be thorough and consider all reasonable lines of enquiry. These should include those that may indicate the suspect's innocence as well as his guilt.

2. *All information collected* during an investigation should be recorded.

3. Any such record should be retained.

4. Any relevant material should be retained.

5. There should be disclosure of relevant material.

It cannot be over-emphasised that all relevant material must be kept in an accessible and durable form, and all relevant information should be recorded contemporaneously or as soon as practicable after the event. All such information should be dated and signed by the officer obtaining it if he is to resist any challenges about its authenticity or his recollection of events in court.

Retaining relevant material

The investigator has to retain all relevant material, defined in the Code of Practice as "material of any kind including information and objects". This is likely to include:

• Details of the offences written in an inspector's notebook.

• Records of conversations, including telephone calls.

• Witness statements, including any earlier drafts.

• Records of interviews.

• Correspondence with expert witnesses, e.g. analysts or engineers.

• Anything casting doubt on the reliability of witnesses or confessions.

All of the relevant material should be retained pending a decision on whether to prosecute and an investigating officer, once he feels that a decision to prosecute is likely, should ensure that all his evidence is collected, recorded and retained in a manner consistent with his obligations and powers under the Health and Safety at Work etc. Act 1974, the Criminal Procedure and Investigations Act 1996 and related legislation. The material must also be retained, once legal proceedings are taken, until the defendant has been acquitted; the case withdrawn; any appeal disposed of; or, where a defendant is found guilty, after release from custody or six months after conviction.

Collecting evidence

In investigating criminal offences, there will be a need to obtain various types of evidence. This may include:

1. Oral evidence.

2. Documentary evidence, including witness statements, transcripts of interviews under caution (PACE interviews) and paper records.

3. Physical or "real" evidence.

4.   Circumstantial evidence.

In order for any evidence collected to be admissible in court, it must comply with established rules of evidence.

Oral evidence
This will be witness evidence given in court by individuals and may include:

- The investigating inspector.

- An employer or employee.

- An independent expert witness.

- A public analyst.

- A photographer or developer of photographs.

Whilst the prosecuting authority will expect to produce witnesses in support of its case, the defence will seek to turn the evidence, or any inconsistencies in it, to its own advantage. It is not unknown for the defence to produce evidence that a state of affairs which is the subject of prosecution has previously been allowed to pass without comment by another inspector. That inspector could be called upon to give evidence for the defence. Whilst this may be no defence to the allegations, it can be most embarrassing for the prosecution.

Documentary evidence
Documentary evidence is information which is recorded. This type of evidence may include:

- *Health and safety policy statements.* The policy statement, especially when compared to the actual practice will enable an inspector to see whether the employer is merely playing lip service to health and safety.

- *Safety instructions.* Examination of any instructions specific to any perceived risk, particularly when investigating an accident, will enable a judgement to be made on the adequacy of the instruction.

- *Risk assessments and method statements.* Examining these types of documents will help to determine, possibly in conjunction with expert advice, whether a dangerous activity and the measures taken to reduce the risk are sufficient to meet the employer's statutory responsibilities.

- *Photographs or videos* of conditions at the time of the alleged offence. Such records, together with any sketches produced at the time of investigation, can provide a far better indication of conditions found than can be conveyed just by the statement of an investigating officer and are therefore an extremely valuable type of evidence.

- *Accident records.* Such records, or the lack of them, will give a useful indication of an employer's historical safety performance.

- Machinery manufacturers' *safety and maintenance advice.* Adherence to such advice will help to show an employer's commitment to safety.

- *Certificates of safety* checks.

- *Instructions* given to specific members of staff, e.g. warnings about conduct in relation to safety or instructions not to use particular pieces of equipment or engage in particular activities.

  The degree of adherence to such instructions will indicate the extent both of the employer's and employee's compliance with their obligations.

- *Sound recordings,* e.g. where noise is alleged to exceed certain criteria.

- *Witness statements.*

- *Contract documents.* Contract documents may show the extent of an employer's activities and should be obtained as a means of deciding who has duties under health and safety law. This may be particularly important where there are a number of occupiers of the same premises or a number of contractors employed on the same area of work.

- *Company documents.* Details of the company's registered office, its directors and a copy of its accounts may all be necessary information, especially in the case of a prosecution. If prosecution of any senior managers is being considered[11] then copies of relevant minutes or other documents relating to their involvement in a particular matter may be required.

- *CCTV recordings.* These may be particularly useful in showing events at the time of occurrence, e.g. a transport accident within the boundary of the employer's premises.

- *Expert witness reports.*

- *Register entries.* Entries in registers, the lack of legally required entries or the complete absence of a required register may be important factors in an investigation, e.g. failure to record an accident known to have occurred or the absence of the particulars of young persons who are employed. If a register is produced for examination, contains relevant information and is retained by the employer, then the inspector should sign and date it below the last entry and/or relevant entries and take copies of the relevant parts of the register.

- *Coroner's report.* In the event of a death, a coroner's inquest usually takes place before any prosecution and may result in the defence making statements or admissions that could be useful to the prosecution. It can therefore be useful to liaise with the coroner's office with regard to the hearing and the evidence submitted. Transcripts of the inquest and copies of relevant documents submitted in evidence can be obtained from the coroner. These may include copies of post-mortem examinations, notes of evidence and any other documents submitted in evidence.[12] The coroner may allow the inspection of these documents without charge. It may be useful for an inspector to attend the inquest to take notes of the proceedings, record relevant information and to form a view about the attitude and degree of co-operation of any witnesses who may be involved in any proceedings with which the inspector may be involved.

---

[11] Under s.37, Health and Safety at Work etc. Act 1974.
[12] Coroners Rules 1984 (S.I. 1984 No. 552), Rule 57 as amended by S.I. 1985 No. 1414.

- *Enforcing letters, enforcement notices and reports of previous local authority investigations* (possibly into similar incidents).

Physical or real evidence
Physical or real evidence is generally a material object, the nature or condition of which is in some way relevant to the alleged offence. This type of evidence may include:

- A piece of equipment.

- An article or substance.

Circumstantial evidence

Such evidence is evidence of a fact not in issue but legally relevant to a fact in issue, e.g. if an employer were charged with exposing an employee to risk to his health and safety by providing him with defective personal protection equipment, the presence on the premises of similar equipment in a like condition would be circumstantial evidence.

Whether a court would allow circumstantial evidence to be heard would depend on its relevance to the facts in issue.

Obtaining the evidence

It is easier if the evidence required by an inspector is obtained voluntarily. However, once someone is aware that he may be prosecuted, he may well withdraw voluntary co-operation and the use of statutory powers will be required. Section 20 of the Health and Safety at Work etc. Act 1974 provides extensive powers to obtain information when investigating breaches of health and safety legislation.[13] Whatever evidence an inspector obtains, he should keep a detailed and chronological record of it, his actions in securing that evidence and, where necessary, the reasons for his actions. He may have to rely on his records in court in the event of a prosecution. If he does so, rest assured that any defence lawyer will do his best to discredit the evidence, the manner in which it has been obtained and the accuracy of the inspector's recollection of events.

---

[13] See Chapter 3, pp.124-126.

It is usual for the originals of documents relied on in court to be produced at the hearing[14] and inspectors should aim to ensure this is done. However, inspectors only have the power to inspect and take copies of documents.[15] If a copy is insufficient for court proceedings and the owner of the document refuses to produce the original, it may be necessary to use the "any other power which is necessary" provision in section 20(2)(m) and seize the document. This power should be used rarely and only as a last resort if the court is to accept that the action was justified and proportionate.

In cases where only the original document will be sufficient, it may not be appropriate or possible to take the original document if it forms part of a register or set of instructions. In these cases the owner or person responsible for the document may agree to produce the original in court. A witness statement to that effect should be taken from that person. If the owner of the document refuses to agree to produce the original, the fact should be noted in the witness statement and the owner requested to sign and date it. If he refuses this request, it may be necessary to use the section 20(2)(m) power. In taking copies of documents, it is sensible to use any on-site facilities that may exist. This allows the person with responsibility for the documents to see what is happening to them and know that there is no attempt to alter or interfere with them in any way. If the responsible person is not present when documents are copied, he should be given a notice identifying the document, stating that it has been taken into the inspector's possession under his section 20 powers and that it will be returned as soon as the copy has been taken. The time the document was taken away and returned should be recorded, together with the reason for copying it.

Occasionally, a situation may arise where an inspector wishes to conduct a search of premises, e.g. where he believes an employer is deliberately concealing evidence likely to reveal a breach of the law (such concealment could possibly form the basis of an obstruction charge).[16] The Health and Safety at Work etc. Act

---

[14] Except in the case of film and tape recordings.
[15] s.20(2)(k), Health and Safety at Work etc. Act 1974.
[16] Although there does not appear to be any specific case law on this issue.

1974 does not provide such specific powers, even though section 20 is wide-ranging in what it allows an inspector to do. Accordingly, any search would have to be by consent, following the provisions of Code B of the Criminal Evidence Codes of Practice.[17] It would be necessary to obtain written consent from a person entitled to grant entry to the premises, e.g. the owner, occupier or manager. Before obtaining that consent, it is necessary to advise the person:

(a) of the purpose of the search;

(b) that they are not obliged to consent; and

(c) that anything seized may be used as evidence.

The Code does not apply to the exercise of statutory powers to enter premises or to inspect plant, equipment or procedures if the exercise of the power is not dependent on the existence of grounds for suspecting that an offence may have been committed, *and the person exercising the power has no reasonable grounds for such suspicion.* This would usually be the case at the start of most routine health and safety inspections and Code B would not apply at that point. However, if an offence is suspected or becomes obvious during a routine inspection, then the provisions of the Code must be followed.[18]

Any searches must be carried out at a reasonable time and only to the extent necessary to achieve the object of the search. The property and privacy of the occupier must be properly considered and a proper search record must be completed containing specific details.[19]

If any evidence has not been properly obtained, e.g. by exceeding the powers available under section 20 or by not following the PACE Code when necessary, a court could rule that the evidence was inadmissible on the grounds that it would unfairly effect the

---

[17] Made under the Police and Criminal Evidence Act 1984.
[18] The Code was altered following the case of *Dudley Metropolitan Borough Council v. Debenhams plc* [1994] 159 J.P. 18, DC in which it had been held that a routine inspection amounted to a search and Code B should have applied from the outset.
[19] Code B, para. 8.1.

fairness of the proceedings. A civil action for damages could follow as a result[20] and, if the matter is serious enough, it might be argued that an abuse of process had occurred.

It is often necessary, once a record has been examined, for further questions to be asked of witnesses as a result of information obtained from the documents. Some examples include:

- Safety instructions may have to be checked to see if employees are aware of them and have been trained in their application.

- Maintenance programmes may need to be followed up to check that what is required has been carried out in practice.

- The results of expert examination of samples or objects may require the interview of experts to secure a full understanding and explanation of their report.

- Where an accident report reveals steps taken to prevent such an incident, it may be necessary to obtain expert advice on whether the steps taken were adequate and "reasonably practicable".

In most cases involving taking statements this will be done by face-to-face contact and the production of a witness statement. There are occasions, however, when this is not possible, e.g. the witness is not local and it is not convenient for any party to deal with the matter locally. In such cases section 20 provides for written information to be obtained.[21] It would then be necessary to write to the witness, sending a form for completion containing a written caution to be signed by the witness, together with a statement form on which his evidence should be written. Such a request for a statement may be accompanied by a specific set of questions relating to the alleged offence, which the witness is invited to answer. The response may then be used as evidence in subsequent legal proceedings.

---

[20] Although s.26 allows an enforcement authority to indemnify inspectors against action.

[21] *London Borough of Wandsworth v. South Western Magistrates' Court* [2003] E.W.H.C. 1158 Admin.

Obtaining computer information

Information is increasingly stored on computers and it may be necessary to rely on this type of evidence. If it is established that information likely to be of value in an investigation is held on computer, the duty holder should be requested to print it out.[22] The inspector should enquire if the computer is working properly. In any prosecution, it would have to be shown:

1. That there are no reasonable grounds for believing that the statement was inaccurate through improper use of the computer.

2. That at all times the computer was working properly or that, if not, any deficiency would not affect the accuracy of the computer-generated document.[23] If there is any doubt about this, the duty holder is unwilling to co-operate or there is uncertainty about the accuracy or reliability of the information, it may be necessary to take along a computer expert[24] to examine the computer and take copies of any required documents. A large amount of material may require an examination which takes the computer out of use for a period of time. The manner in which the computer is examined must minimise the amount of down time so that, if required, the action of the inspector can be shown to be proportionate and necessary.

Evidence as to the operation of a computer may be crucial to the prosecution or defence where it contains information on which they rely. A case may fall if such evidence is not made available.[25]

Taking possession of articles or substances

An inspector has the power to take possession of and detain any article or substance which appears to have caused or be likely to cause danger to health and safety, in order to examine it, ensure it is not tampered with or to ensure that it is available as evidence in any legal proceedings.

---

[22] Relying on s.20(2)(m) powers.
[23] Police and Criminal Evidence Act 1984, s.69.
[24] Using s.20(2)(c) powers.
[25] Evidence can be provided by a certificate in the form shown in Sch. 3, Part II to the Police and Criminal Evidence Act 1984. It can also be given by any person familiar with the operation of the computer, *R. v. Shephard* [1983] A.C. 380.

If an article or substance is readily removable by the inspector, this power is straightforward. If, because of its weight, size or other circumstances preventing its immediate removal it has to remain in position for a period of time, the duty holder or responsible person should be given written notice that it has been taken possession of and detained, and cautioned that if he attempts to remove or interfere with it he may be committing a further offence. In such cases, particularly if the inspector is unaccompanied and has no colleague to maintain watch over the item while arrangements are made for its removal, it would be sensible to make any relevant drawings, notes or take photographs relating to the position or condition of the item.

At the same time, an inspector has the power to direct that the premises, or any part of them, or anything in them be left undisturbed for so long as is reasonably necessary for the purpose of any examination or investigation.[26] The power might be used in cases where:

- The condition of the site or the layout and condition of equipment may be essential evidence.

- The conditions following an accident, including any damage that may have been caused, may be crucial in deciding, together with other evidence, what actually happened.

- An expert investigation may be required into an incident, including the examination of equipment.

- It may be necessary to return with specialist sampling, analytical and/or measurement facilities.

Although it may be possible to agree informally with the employer or person in control to leave an area or facilities undisturbed, if a prosecution is likely to follow it is preferable to issue a formal notice requiring things to be left undisturbed. That notice should refer to the inspector's specific powers[27] and, where practicable, give a time during which the evidence must not be disturbed. If

---

[26] s.20(2)(e).
[27] *ibid.*

necessary that time period can always be extended. The employer, particularly if he believes that legal proceedings may follow, may want to obtain his own expert report. He should be allowed to do so. However, the inspector should not rely on that report rather than his own expert report as it is by no means unknown for experts to disagree! In any case, the employer's report is unlikely to be released to the inspector as he may be entitled to withhold it on the grounds of *legal professional privilege*.[28] The scientific content of a report should not attract professional privilege: it would have to be disclosed by a defendant in criminal proceedings if he wished to rely on it in a prosecution.

When using the power to take possession of an article or substance which appears to have caused or be likely to cause danger to health and safety, the inspector must leave a notice with a responsible person at the premises or, if this is impracticable, fix it in a conspicuous position. Before taking possession of an article or substance, the inspector must, if practicable, take a sample and give a portion of it marked in a manner sufficient to identify it, to a responsible person on the premises.[29]

Sampling
An inspector has the power to take samples of any articles or substances, including atmospheric samples.[30] In cases where special equipment has to be used, if necessary it should have a current calibration certificate. Where it is equipment which should be calibrated before each use, evidence that it has been calibrated must be available at any court hearing. Where a sample has been taken, it should be divided into three parts, each properly labelled to identify them, and one part should be given to the occupier, one submitted for analysis and one retained by the inspector. Each item sampled must be identified by a label containing a sufficient description to identify it, a number if there is more than one sample, the premises from which it was obtained, the date and time of sampling, and the name and position of the person taking the sample.

---

[28]   s.20(8). See p.170.
[29]   s.20(6).
[30]   s.20(2)(g).

If it is not practicable to divide a sample into three parts, the potential defendant should be advised that a prosecution might occur and be invited to take his own sample if he so wishes. If an item is likely to be damaged or destroyed as a result of any test to which it is submitted,[31] the potential defendant should be notified before the test is conducted and given the opportunity to be present during the test. Persons who were present at the premises and have responsibilities in relation to those premises are entitled to request that they be present when any article or substance is being examined, tested or analysed.[32]

As far as possible there should be retained sufficient of the article or substance to allow independent examination. If this is not possible, other steps should be taken to enable the results to be produced to the court, e.g. photographing or video recording the examination and keeping detailed notes and records of all results.

In the case of samples taken and submitted for analysis, as with all other exhibits, there must be an unbroken chain of evidence demonstrating that each sample has been kept safely and treated appropriately at all stages of handling and examination up to the court hearing, e.g. the statement from the laboratory which analysed a sample must record the unique labelling attached to the sample by the inspector who took it, labelled it and sent it for analysis. A properly documented system must be in place, not just for samples, but for all articles or substances which may be used in evidence, whether or not they have been subjected to a formal sampling procedure. It is not always necessary to produce the sample in court,[33] indeed some samples may have deteriorated by then, but it may be useful to do so in cases where this helps the court in its deliberations.

Analytical and test results
If laboratory or other scientific tests are carried out, evidence must be given about that work by the person who carried it out or by the person supervising or controlling the work.[34] If a supervisor

---

[31] s.20(2)(h) recognises this possibility.
[32] s.20(4).
[33] *R. v. Orrell* [1972] Crim LR 313.
[34] *Stone and Sons v. Pugh* [1948] 2 All ER 818; *R. v. Kershberg* [1976] RTR 526; *English Exporters (U.K.) Ltd. v. Eldonwall Ltd.* [1973] Ch 415 at 420 F.

attends court, they must be able to understand, describe and answer questions relating to the work of the person under their supervision. It is not good enough just to relay things said by the individual who conducted the examination simply because that person may be an expert in their field.[35] It is also worth bearing in mind that a defendant may ask for a copy of, or the opportunity to examine the record of, any observation, text, calculation or other procedure on which the examiner's findings are based. This may result in further detailed questioning in court. It is, therefore, almost always best to have the person who conducted the examination giving first-hand evidence in court.

The use of legal professional privilege

An inspector cannot require the production of documents from a person who would be entitled to withhold them on the grounds of legal professional privilege.[36] This extends to counsel, solicitors, their clerks and agents. It includes reports, statements and information created during a lawyer/client relationship. The law is fairly complex on this issue[37] but it is the client who is entitled to waive the privilege and produce or use a document covered by it if they so wish. An example of a case where an inspector might wish to obtain a privileged document could be an engineering report obtained by a defendant with a view to deciding whether to contest legal proceedings. He would be unable to do so unless the defendant chose to waive his privilege (although presumably he would only do so if it benefited him!).

IMPACT OF THE REGULATION OF INVESTIGATORY POWERS ACT 2000

This legislation was enacted partly as a result of the surge in communications technology, and to protect human rights following the introduction of the Human Rights Act 1998. Section 6 of the 1998 Act is arguably the most significant, making it unlawful for a public authority to act in a manner which is incompatible with a Convention right. In relation to health and safety, Article 2 deals

---

[35] *R. v. Abadom* [1983] 1 WLR 126.
[36] s.20(8).
[37] See a useful summary in *Collecting physical evidence – obtaining evidence using section 20 powers,* Health and Safety Executive, September 11, 2003, HSE.

with the right to life and applies to cases involving a risk of death, so one effect of this Article is that local authorities may be under a general duty to protect people from life-threatening situations. Article 8 stipulates the right to respect for private and family life and could be invoked in some circumstances where health and safety investigations are conducted.

When investigating health and safety complaints, local authorities may carry out monitoring of employment conditions and the activities of individuals. Whether Article 8 is contravened will depend on whether surveillance is being undertaken. This is defined in the Regulation of Investigatory Powers Act 2000[38] as including "monitoring, observing or listening to persons ... their conversations or their activities or communications". It also includes "recording anything monitored, observed or listened to in the course of surveillance". Activities not authorised under the Act are unlawful under section 6 of the Human Rights Act 1998 and accordingly any evidence collected in an unauthorised way may be inadmissible in a court of law.

Part II of the 2000 Act may apply to local authority investigations, although the Act is intended primarily to be associated with police, security services and HM Customs and Excise activities. The Act applies to directed surveillance, intrusive surveillance, and the conduct and use of covert human intelligence sources.[39]

"Directed surveillance" is defined as covert[40] but not intrusive when undertaken as a specific investigation or operation and carried out in such a way as to make it likely that private information is obtained about a person. "Private information" in relation to a person includes any information relating to his private or family life.[41] "Surveillance" includes monitoring, observing or listening to persons and any recordings made as a result of such activities.[42]

Investigations which involve the monitoring of an employee's activities, such as the way he drives a fork lift truck or handles

---

[38] s.48.
[39] s.26(1).
[40] s.26(2).
[41] s.26(10).
[42] s.48(2).

awkward or heavy items, might be included in this definition when carried out covertly, perhaps using photographic equipment. Whether the evidence obtained is private is not clear. It might be argued that merely confirming the existence of unsatisfactory practices is not intrusive. Whether such evidence is information relating to family or private life would have to be determined by the courts.[43]

"Directed surveillance" must be authorised in each case by a "designated person", e.g. an Assistant Chief Officer or an officer responsible for the management of an investigation.[44] Authorisation must be obtained for each investigation.

"Intrusive surveillance" is covert surveillance if it is carried out in relation to anything taking place on any residential premises or in any private vehicle; and involves the presence of an individual on the premises or in the vehicle or is carried out by the means of a surveillance device.[45]

The Act also says that surveillance is not intrusive if carried out without that device being present on the premises or vehicle, unless it consistently provides information of the same quality and detail as might be expected if the device was actually present on the premises or vehicle.[46]

It is important to distinguish between the two types of surveillance, as intrusive surveillance by local authorities is not authorised by the 2000 Act. Whether a video camera operated in the investigation of a health and safety issue represents surveillance for the purposes of the Act is not clear-cut but, if it were deemed intrusive, the evidence obtained by it may not be authorised by the Act and therefore may be inadmissible in court. The covert use of such a camera could well be intrusive surveillance, the undertaking of which by local authorities is prohibited. As the Regulation of Investigatory Powers Act 2000 was primarily intended to regulate the activities of services more dependent on surveillance, it is not

---

[43]   There does not appear to be case law on this issue.
[44]   Regulation of Investigatory Powers (Prescription of Officers, Ranks and Positions) Order 2000, S.I. 2000 No. 2417.
[45]   s.26(3).
[46]   s.26(5).

clear whether the type of regulatory monitoring undertaken by local authorities was properly considered.

INTERVIEWING WITNESSES AND SUSPECTS

Collecting witness statements

Witness statements contain the evidence of someone spoken to about a particular matter and the signature of the person to confirm the truth of their evidence. The statement should only contain information about things they have seen, not what they surmise or have heard someone else say. However, anything which helps develop an investigation or corroborates other information should be recorded.

There are two types of statement:

1. Those obtained *voluntarily* in compliance with section 9 of the Criminal Justice and Public Order Act 1994. These do not rely on the power of compulsion in the Health and Safety at Work etc. Act 1974, they are likely to be taken during the earlier stages of an investigation and are for information gathering generally. Such statements are admissible in court without requiring the attendance of the witness, subject to certain conditions:

    (a) the statement is signed by the person making it;

    (b) it contains a statement by the person making it that it is true to the best of his knowledge and belief and that it was made knowing that, if tendered in evidence, he would be liable to prosecution if he wilfully stated anything in it which he knew to be false or did not believe to be true;

    (c) a copy of the statement is served on the other parties to the case before the court hearing;

    (d) none of the parties involved objects to the statement being tendered in evidence. It has been known, however, for defence lawyers to accept such a statement and then on the day of the hearing demand that the witness give evidence, a tactic which can delay and cause some stress and confusion for the other party (presumably deliberately).[47]

2. Those obtained *compulsorily* under the Health and Safety at Work etc. Act 1974.[48] Section 20 provides the power to require any person whom the inspector has reasonable cause to believe will be able to provide information relevant to an examination or investigation, to answer questions and sign a declaration of the truth of the answers. This type of statement may include admissions of liability. It should be borne in mind that, in requiring information under the Health and Safety at Work etc. Act 1974, no answer given is admissible in evidence against the person providing it, or their spouse.[49]

Witness statements should be obtained as soon as possible after the events to which they relate so that:

(a) they are still fresh in the witness's mind; and

(b) the evidence is recorded before the witness has a chance to change his mind about events or discuss it with anyone else.

By taking statements early on there is less chance of the defence challenging its accuracy.

Interviewing

The Police and Criminal Evidence Act 1984 (PACE) and its associated Codes of Practice govern, amongst other things, the interviewing of suspects. Section 67(9) of the Act requires that persons other than police officers who are investigating alleged offences shall have regard to the relevant provisions of the Codes. The Codes must be followed.[50] Code C deals with the interviewing of suspects and, if the interview is conducted using tape recorders, then Code E also applies.

---

[47] Other provisions under s.9 must be followed relating to people under the age of 18 years, where the witness cannot read, and where the statement refers to any documents as exhibits.

[48] s.20(2)(j).

[49] s.20(8). This is also supported by Art. 6 of the European Convention on Human Rights.

[50] *R. v. Elson, The Times,* June 30, 1994.

A PACE interview is defined as " … the questioning of a person regarding his involvement or suspected involvement in a criminal offence". Although much of Code C is relevant only to police officers,[51] those elements relevant to the work of local authority investigators must be taken into account. Interviews under caution should always be tape-recorded, although this is still not universal practice amongst local authorities.

Who should be interviewed?

The people to be interviewed may include company directors, board members (particularly those with defined health and safety responsibilities), managers, supervisors, employees, the self-employed and members of the public or visitors to a place of work. Interviews should involve people with knowledge of the circumstances leading to a particular incident or safety issue, the systems of work, the safety policy, the supervision, training arrangements, etc., indeed anything which may produce information relevant to the matter under investigation. Some of those people may have committed an offence. If this is suspected, they should be formally interviewed under the provisions of the Police and Criminal Evidence Act 1984.

The purpose of the interview should be to obtain the best possible evidence.

The principle reasons for a PACE interview are:

1. To provide the suspect with the opportunity of putting forward their own version of events.

2. For the investigator to obtain further evidence.[52]

3. To clarify any matters that may be necessary.

4. To help decide the appropriate enforcement action.

Preparing for the interview

Preparing for the interview is most important and a copy of the

---

[51] e.g. the detention and treatment of people in custody.
[52] Although the investigating officer can also obtain information using his powers under s.20(2)(j) of the Health and Safety at Work etc. Act 1974.

relevant Code of Practice should be available for use by the suspect during the interview. The interviewer should be trained in the law and the interview process. A lot of defendants are convicted on the basis of the information and admissions made at interview and it is therefore important that the process is conducted in accordance with the rules laid down.

It is good practice to write to the suspect inviting him for interview and stating the nature of the alleged offences. This will also give the suspect time to prepare for the interview, including trying to defend his position if he believes he may have committed an offence. The investigating officer should be prepared for this. The officer may wish to prepare a set of relevant questions relating to an alleged offence to aid the interview process. If so, he may also wish to consider any likely defences the interviewee may produce so that he can develop his questioning accordingly.

When inviting a witness or suspect to an interview it would be good practice to include the following:

- The nature of the offence under investigation.

- The right to legal representation.

- The voluntary nature of the interview.

- The inclusion of a number of alternative dates and times for the interview or confirmation of any verbally agreed arrangements.

- Question whether the suspect requires the attendance of an interpreter. If the ethnicity of a non-English speaking person is known, it may be appropriate to have the letter translated into the suspect's own language.

- If the interview is to be taped then this should be made clear.

Before interviewing the suspect, the investigating officer should consider the nature of the offences he is investigating by identifying the "facts in issue", i.e. the identity of the offender and the availability of any statutory defences. He can then decide on the complexity or otherwise of the case and prepare accordingly.

The interview should concentrate on a series of relevant questions to be put to the witness/suspect. The detailed questions will depend on the particular circumstances being investigated but most interviews should probably seek to clarify the following:

1. Who is responsible for the actions or omissions giving rise to the alleged offence and/or the person who is legally accountable. An employee may be accountable for a failure to protect his own safety but his employer may also be accountable for not ensuring that his employee complied with specified safety instructions.

2. Whether the suspect is actually a suspect or merely a witness. It has been established that the PACE Codes of Practice do not apply to an employee in connection with an offence where the employer itself is likely to be prosecuted and the employee is a *witness rather than a suspect*.[53] Where the interviewee is an employee, it may be appropriate to allow him an opportunity to consult his employer before answering questions, especially in cases where the employee may be subject to proceedings. This might not be appropriate, however, where collusion is suspected.

3. The nature of the facts in the case. This should for example result in the clarification of events leading to an alleged offence, the actions of individuals involved and any action subsequently taken.

4. If the results to date of an investigation have been assembled, the interviewer will be able to ask questions to fill any gaps in his information in order to provide a detailed and chronological sequence of events leading to the alleged commission of the offence.

5. The interviewer should identify any potential defences available to the suspect so that he can prepare a series of questions designed to explore the strength of those defences in the event of a prosecution.

---

[53]   *Walkers Snack Foods Ltd. v. Coventry City Council* [1998] 3 All ER 163.

Conducting the interview

Once a witness becomes a *suspect*, e.g. during the course of a voluntary interview[54] or as a result of a compulsory statement,[55] a caution must be issued in accordance with the Police and Criminal Evidence Act 1984 and a formal interview conducted in accordance with the Act and its Codes of Practice. To ensure fairness, it is better to conduct this as a separate interview, at a later date, and with a different officer if incriminating admissions have been obtained during the previous interview.

If a person who is interviewed prior to being charged with an offence fails to mention anything which he later relies on in his defence, the court can draw such inferences from the failure as appears proper.[56]

The interview of a person suspected of an offence has to be fair[57] and any record of the interview must be accurate and fair. A court can refuse to allow evidence if it appears that *having regard to all the circumstances, including the circumstances in which the evidence was obtained,* the admission of the evidence would have such an adverse effect on the proceedings that the court ought not to admit it.[58] The presumption in the Police and Criminal Evidence Act 1984 and the Codes of Practice is that the interviewee knows why he is being interviewed and the nature of the offence involved. Interviews are not trials and do not require the type of aggressive questioning portrayed in some television programmes. Indeed, this is prohibited by section 76 of the Police and Criminal Evidence Act 1984. If the interviewing officer suspects or knows that a suspect is lying, he should use probing questions to reveal the truth and give the suspect or witness the chance to change or clarify his evidence.

In the case of taped interviews, the whole of the interview should be recorded and nothing said whilst the tape is switched off. The parties to the interview should be introduced on the tape and the

---

[54] Under the Criminal Justice and Public Order Act 1994.
[55] Under s.20 of the Health and Safety at Work etc. Act 1974.
[56] s.34, Criminal Justice and Public Order Act 1994.
[57] s.78, Police and Criminal Evidence Act 1984.
[58] *ibid.,* s.78(1).

suspect given a copy of the tape on completion of the interview. The practices should follow as far as practicable the terms of PACE Code C:[59]

"(a) an accurate record must be made of each interview with a person suspected of an offence...;

(b) the record must state the place of the interview, the time it begins and ends, the time the record is made (if different), any breaks in the interview and the names of all those present...;

(c) the record must be made in the course of the interview unless in the investigating officer's view this would not be practicable or would interfere with the conduct of the interview... ."

Where the interview is tape recorded, the requirements of PACE Code E must also be followed.

If an interview record is not made during the course of the interview, it must be made as soon as practicable after completion. Written interviews must be timed and signed by the maker. If an interview record is not completed in the course of the interview, the reason must be recorded. Unless it is impracticable, the person interviewed must be given the opportunity to read the interview record and sign it as correct or indicate in which respects he considers it inaccurate. Any refusal by a person to sign an interview record when asked to do so must be recorded. A written record must also be made of any unsolicited comments made by a suspect which are outside the context of the interview but which might be relevant to the offence. That record must be timed and signed by the maker and, where practicable, the person concerned must be given the opportunity to read the record, sign it as to its correctness or indicate in what respects it is considered inaccurate. Any refusal to sign it must be recorded.

In some cases where it is not practicable to conduct a face-to-face interview, e.g. because the witness or suspect is out of the country

---

[59]   para. 11.5.

for some time or is otherwise not readily available, it is quite common for interviews to be conducted by a "written interview". These are not covered by the PACE Codes but generally consist of a letter containing a suitable form of caution, together with a series of questions to which the recipient is invited to respond. The response can then be admitted as evidence.

Failing to administer the caution
Failure to administer a caution before interview is usually considered by the courts to be a significant breach of the PACE Codes and it is likely that the evidence will be excluded under the Police and Criminal Evidence Act 1984.[60] The main issue will be whether there has been such an adverse effect on the fairness of the proceedings that the evidence ought to be excluded.

Failure to offer the opportunity to obtain legal advice
If a person is arrested and held in a police station, or interviewed there, and is not advised of his right to legal advice before interview, the courts may take a dim view of the situation and use their discretion to exclude the evidence obtained.[61] A person interviewed in connection with an offence under health and safety legislation may also be entitled to take legal advice before answering any questions.

Location of the interview
The requirements of the Act and the PACE Codes can be onerous. It is therefore best if interviews are carried out in an environment that will be free from interruptions, have facilities for toilet and refreshment breaks, and provide an atmosphere in which the interviewee feels relatively comfortable. Not all people are fully aware of the implications of the interview until they are cautioned. Their demeanour may then change and they might become aggressive at some point. Having regard to these points, interviews are usually best undertaken in a police station or local authority offices where assistance is at hand if needed.

Rooms used for interviews should be well lit and ventilated, with

---

[60] s.78. See also *R. v. Sparks* [1991] Crim. L.R. 128.
[61] s.78, Police and Criminal Evidence Act 1984.

a comfortable temperature maintained. A table and chairs should be provided and laid out so that the interviewing officers and the interviewee and any legal or other representative sit on opposite sides of the table.

Officers' duties
Two officers are usually involved in interviews, with one taking the lead role in questioning and, the other ensuring that the proceedings follow the PACE Code requirements, operating any tape recorders or taking down written statements, dealing with any exhibits and asking questions or prompting his colleague in respect of anything he may have omitted. It is often useful for the lead officer to invite any additional questions or comments from him.

Questions should be unambiguous so that the interviewee is clear about what is being asked. Many officers feel that tape-recorded interviews are preferable as there can be absolutely no doubt about what has taken place at an interview and, if necessary, written transcripts presented in court can be challenged by reference to the tape.

Beginning the interview
The way in which the interview will be conducted should be discussed, even though this should have been made clear in the written invitation to the interview. The interviewing officer should introduce all the participants and state the location, date and time of commencement. The interviewee should be informed that he is not under arrest, can terminate the interview at any time and may have a legal representative present if he wishes. In the absence of any legal representative, the consent of the interviewee should be obtained to continue the interview, with the fact recorded.

In the case of the Health and Safety at Work etc. Act 1974, offences may be committed by both the employer and/or employee. In such cases, if the officer suspects that both are involved in an offence, each person should be interviewed separately and they should not be represented by the same solicitor if that would involve a conflict of interest, e.g. giving evidence leading to the prosecution of the other party.

Where interviews are to be taped, the interviewee should be asked if he agrees and the agreement or otherwise recorded. If a written record is to be made, the interviewee should be asked to speak clearly and at a pace which enables a verbatim record to be taken. In the case of tape-recorded interviews, written information should be provided explaining the use to which the recordings will be put, and all information should be properly logged and accounted for using standardised forms.[62]

The interview should then follow a basic structure:

1. *Obtaining personal information* – name and address, any trading name, company registered office, partnership details.

2. *Stating the reasons for the interview*, e.g. the nature of the offence being investigated.

3. *Administering the caution.* This is only required once there is sufficient information to suspect that an offence has been committed and must be given in the following terms: *"You do not have to say anything. But it may harm your defence if you do not mention when questioned something which you later rely* on *in court. Anything you do say may be given in evidence."*[63] The time of the caution should be recorded and the interviewee asked if he understands it. If he does not, it should be explained in simple language. Once the officer is clear that the caution is understood, he should record the fact. Interviewing officers will be familiar with, and understand, the caution but just because a defendant says he understands it does not necessarily mean that is so. He may be frightened, have no idea what the caution means even though he has heard the words before, he could be drunk or not understand the English language well enough. Ask him to explain the caution in his own words and, if it is clear he does not fully understand it, it may be wise to suspend the interview to allow him to obtain the services of a solicitor. If necessary an

---

[62]   Valuable information and advice about tape recording statements is contained in A *Practical Guide to Criminal Investigations for Local Government Officers*, Bowles, Andy, Chadwick House Group Limited.

[63]   PACE Code C, para. 10.4.

interpreter[64] should be used, but the need for one ought to be determined prior to the interview where possible, as obtaining one can be time-consuming (and costly).

Deciding when to give a caution can cause problems when an officer is trying to establish the facts in a case and wants to get as much information as possible before administering the caution. However, "A person whom there are grounds to suspect of an offence *must be cautioned* before any questions about it ... are put to him regarding his involvement or suspected involvement in that offence *if his answer or silence ... may be given in evidence to a court in a prosecution.*"[65] Nevertheless, there is no benefit in cautioning someone too soon. To do so runs the risk of the individual relying on his right of silence and not answering questions or providing information.

Questioning without cautioning the individual is quite legitimate if the officer is seeking information to determine whether an offence has been committed. This applies to an investigation at the scene of a possible offence as well as at a formal interview. Cautioning someone too soon may result in them refusing to answer questions, making it difficult to decide if an offence has occurred.

It is important to remember that an interview is a crucial means of obtaining evidence and the most useful information is likely to be obtained if the interviewee is treated properly in accordance with the provisions of the relevant parts of the Codes of Practice.[66] This includes treating him with courtesy and consideration, even if this

---

[64] The use of an interpreter should be free by virtue of Art. 6(3)(e) of the European Convention on Human Rights.

[65] PACE Code C, para. 10.1.

[66] If all elements of the interview process are not conducted correctly there may be a breach of Art. 6 of the European Convention on Human Rights – the right to a fair trial. For information on the use of interpreters and the relationship to the fair trial provisions of the Human Rights Convention, see *R. (Boskurt) v. Thames Magistrates' Court* [2001] E.W.H.C. 400 Admin; [2002] R.T.R. 15. Anyone charged with a criminal offence has the right, where they cannot understand or speak the language used in court, to the free services of an interpreter regardless of their financial position: Art. 6(3)(e) of the European Convention on Human Rights. See also *Luedicke, Belkacem and Koc v. Germany* (1978) 2 EHRR 149.

is not returned. If contradictory evidence is given, this should be mentioned and the interviewee asked to clarify which version is correct. If exhibits are produced, e.g. photographs, samples, or items previously removed from premises under the officer's statutory powers, they must be described and referred to by their labelling.

Stopping or suspending the interview

Breaks will be required during an interview, especially if the interview is lengthy, the interviewee becomes tired or distressed, or it is interrupted for any reason. In any case, Code of Practice C[67] requires that there be a break from the interview for refreshment, i.e. normal meal breaks and at least 15 minutes at two-hourly intervals. It will be necessary, particularly during a lengthy interview and especially after a break in proceedings, to remind the interviewee where he has already been cautioned, that he is still under caution.

Suspects should not be left alone during an interview and specific requirements deal with the arrangements for handling tapes in the case of a taped interview.[68] If interviews with suspects are tape-recorded, a court may exclude evidence of the interview if a relevant requirement of the Code is not followed.[69]

Once the officer believes that a prosecution should be brought *and there is sufficient evidence for it to succeed,* he should ask whether the suspect has anything further to say and, if not, must cease to question him. However, in practice, an officer will often not be in this position at the time of an interview because the bulk of the evidence likely to be used is not confession evidence.

Juveniles, mentally disordered and mentally handicapped people

A juvenile or a person who is mentally disordered or mentally handicapped, whether a suspect or not, must not be interviewed or asked to provide or sign a written statement in the absence of an

---

[67]　para. 11.16.
[68]　See Code C.
[69]　For detailed information on the tape recording of interviews, see PACE Code E; HSE Enforcement Guide *Collecting Witness Evidence – Questioning of Suspects*; and *A Practical Guide to Criminal Investigations for Local Government Officers*, Bowles, Andy, Chadwick House Group Limited.

appropriate adult.[70] An appropriate adult for a juvenile would be a parent or guardian, or if that is not possible another responsible adult not employed by the enforcing authority.[71] An appropriate adult for someone mentally disordered or handicapped would be a relative, guardian or other person responsible for care, or someone with experience of dealing with such people.[72] The appropriate adult should not be someone suspected of involvement in the offence or a witness and should be able to communicate effectively with the person being interviewed.

Inferences from silence
The Criminal Justice and Public Order Act 1994[73] provides that a court, in determining whether the defendant is guilty of the offence charged, or in determining whether there is a case to answer, may draw such inferences as appear proper from the evidence of silence in certain circumstances.[74] An inference can only be drawn where an interview under caution takes place and then only subject to a number of conditions set out by the Court of Appeal.[75]

Concluding the interview
When concluding the interview, the interviewee should be informed that he can clarify or add to anything he has already said. The officer should advise that the interview is being suspended or terminated and the time stated. He may also wish to have time to consider the information provided before coming to a decision on the course of action to be taken. In this event, he may wish to advise that he is terminating the interview at that time and will contact the person if he wishes to re-interview him at a later date, or inform him of his decision once his investigations are complete.

The interviewee should then be given the opportunity to read through the written record and identify any inaccuracies. He should

---

[70] PACE Code C, para. 11.13.
[71] PACE Code C, para. 1.7(a).
[72] PACE Code C, para. 1.7(b).
[73] s.34.
[74] Failing to mention any fact relied on in the defence when questioned under caution, or on being charged or officially informed that they might be charged.
[75] *R. v. Argent* [1997] 2 Cr. App. R. 27. See also a useful explanation in *Collecting witness evidence – inferences from silence*, HSE Enforcement Guide (England and Wales), updated September 2003.

then be asked to sign the record, as should his legal representative if present. If he refuses to do so, the fact should be recorded.

Transcripts and summary of interviews
Where legal proceedings are to take place following an interview, the Home Office requires that a written record be made.[76] This can be made by a typist or secretary but the interviewing officer will have to ensure that the record is accurate. A written record of a tape-recorded interview can be produced as a summary or a transcript. The verbatim account contained in a transcript leaves no doubt as to what has been said and by whom and should invariably be produced in the case of short interviews. In the case of longer interviews, a summary may be more appropriate, although this must be a fair reflection of the interview. In the case of a tape-recorded interview, the complete tape will, of course, be available for examination, if required, in court proceedings.

An interpreter should be provided if the interviewee has difficulty in understanding English, the interviewing officer cannot speak his language, and/or the person wishes an interpreter to be present.[77] The interviewing officer must ensure the interpreter makes a written note of the interview at the time it occurs and in the language of the interviewee. The interpreter also has to certify the accuracy of the record made. Sufficient time must be allowed by the interviewing officer for the interpreter to record the questions and answers, and the officer should also take particular care to ensure that the interpreter fully understands the questions. The interviewing officer should ask the questions, allow time for the interpreter to repeat them in the interviewee's own language and the interpreter should write down the reply in the interviewee's language and then translate it for the officer. The interviewee must be given the opportunity to read the record made in his own language and correct any inaccuracies. Because of the complexities of this process, it may be best to tape record such interviews. It is also preferable if the interpreter is properly qualified, with a Diploma in Public Service Interpreting and is also listed on the National Register of Public Service Interpreters.

---

[76]   Home Office Circ. 26/1995.
[77]   PACE Code C, para. 13(1).

Health and safety for interviewing officers

Officers have responsibility for their own safety, as do their employing authorities. Although interviews are usually comparatively safe affairs, it is worth bearing in mind that some interviewees, especially if they know themselves to be guilty of an offence, may occasionally take an aggressive and threatening attitude. A risk assessment[78] should therefore be carried out with a view to reducing the risk of violence. This should produce a number of preventative steps to minimise the risk of personal injury, including:

- Ensuring the presence of more than one officer at all times.

- Using a venue for the interview which is suitable for the purpose, e.g. located at a police station or local authority offices where help is readily available.

- Separating witnesses or suspects from the interviewing officers by suitably sized tables, the provision of panic alarms and positioning officers nearest to the exit of a room.

- Training officers in interview techniques designed to limit the risk of aggressive responses from those being interviewed. Liaison with the police may be especially useful in matters of this kind.

- Anything capable of being used as a weapon should not be allowed in the interview room.

- If required, the use of self-defence courses or any other steps reasonably necessary to enable an officer to respond to the risk of personal violence.

Preparing evidence for court

Exhibits

Usually documents and real evidence will have to be presented in court by witnesses. Accordingly, all documents or objects to be used as evidence will have to be produced as an exhibit by a witness.

---

[78] Management of Health and Safety at Work Regulations 1999. S.I. 1999 No. 3242.

All exhibits will have to be identified by labels or another mark, signed by the person making the statement referring to it. The statement must describe the exhibit sufficiently to identify it. The identification should be made using the initials of the person who will produce it, followed by a consecutive number, e.g. CNP 1. Examples of where exhibits are appropriate include:

- A photograph or sketch of premises should be referred to in a statement by the person who took the photograph or produced the sketch.

- A record of a taped interview should be an exhibit and referred to by the inspector who carried out the PACE interview in his formal statement.

- If any codes of practice or other form of recognised guidance are to be relied on, they should be exhibited in a statement by the person, usually an expert, who will produce them in court.

- Any samples/items should be referred to in a statement by the person who took them and the person who conducted any tests, e.g. an analyst or engineer.

Any exhibits must be kept safely and there must be a clearly identifiable audit trail showing how they were dealt with from the moment they were obtained to the moment they are presented in evidence. This is required because it may be necessary to show that the exhibit presented is that referred to by the witness in his statement, or that it has not been improperly tampered with whilst held pending court proceedings.

This means that each person handling an exhibit must produce a statement identifying it, saying when and where they received it, from whom and for what purpose, and to whom they passed it and when they did so.

Documents
Relevant documents will often have been found in the possession of the defendant and taken by the inspector for use as evidence.[79]

---

[79]   Using s.20 powers under the Health and Safety at Work etc. Act 1974.

The statement should explain how they came to be in the inspector's possession and the person who gave the document should be asked to provide a statement describing it and its contents.

It is important that the validity of company documents such as health and safety reports, policy statements and maintenance records can be relied on by the prosecution at the time of the court case. Accordingly, a director or company secretary should be asked to confirm their validity in a witness statement, or in a PACE interview if they are a suspect.

Subject to certain conditions, a statement in a document is admissible in criminal proceedings as evidence of any fact of which direct oral evidence would be admissible.

Evidence of previous character
Any evidence of the past safety record of a defendant may be relevant to the offence being prosecuted as well as to the court in sentencing following a conviction. Any evidence of prior enforcement action or convictions should be obtained and steps taken prior to the hearing to allow its use. Although evidence of previous convictions is not usually admissible, it may be produced to counter evidence that the defendant was not aware of the danger[80] or had made an innocent mistake.[81]

If a defendant has a record of previous convictions, the defence should be informed that the information will be put before the court.[82] The defence can then make representations about any disputed facts.[83] In such cases the convicting court should be asked to provide a Certificate of Conviction.[84]

In cases where letters or notices have previously been sent to the defendant, the person with that direct knowledge should make a statement detailing the action taken. If this is not possible, e.g. because the officer has moved on, the investigating officer should

---

[80]   *R. v. Bond* [1906] 2 KB 389.
[81]   *R. v. Rance* (1975) 62 Cr. App. 118.
[82]   Under s.104, Magistrates' Courts Act 1980. This information can only be used in rebuttal or after a finding of guilt/admission during sentencing.
[83]   *R. v. Sargeant* (1974) 60 Cr. App. R. 74.
[84]   s.73, Police and Criminal Evidence Act 1984.

produce the documents in a statement, stating where they came from.

Identifying defendants

This can be rather complicated as proceedings are not often taken against an individual alone.

Local authorities will most commonly take proceedings against one or more of the following:

- Companies.

- Partnerships.

- Limited liability partnerships.

- Individuals using a business name.

- An individual in their own name.

The prosecution must correctly identify itself and the defendant in the information, summons or warrant setting out the allegation. The defendant is also entitled to know who is accusing him and that the accuser has the necessary authority to prosecute him. The prosecution must correctly identify the defendant, as amendments cannot usually be made to an information or summons and it will be necessary to re-issue the document, with the attendant delays. Amendments which are merely technical will usually be allowed, but not if they refer to a different offence or are major changes.

Companies

It may be the case that the person being prosecuted operates from commercial premises and the prosecuting authority incorrectly names the company or an individual. This can be a particular problem where a company is one of a group of companies. An example might be where a group of companies is known by a trading name, the staff are employed by another company and the land and buildings in which they work are held by yet another company. In many cases, a company may have no employees.

It is not unknown for a prosecuting authority to proceed against the wrong organisation and not to find this out until they get to court.

A defendant is under no obligation to tell the prosecutor he has got things wrong and may simply challenge the prosecution in court on the basis that he is not the correct defendant. However, deliberately failing to give the correct information needed to identify a defendant may itself be an offence.[85]

A simple error, e.g. a mistake in the defendant's first name where he has otherwise been correctly identified, will not result in the prosecution failing.[86] The court has discretion to decide whether to allow an amendment. However, if the wrong party altogether is brought before the court as a result of an error, then no amendment is possible.[87]

Where proceedings are to be taken against a registered company,[88] the correct details to enable the accurate service of the information and summons can be obtained from a search of the Companies Register. This will provide details[89] of the correct name of the company, the company number, its registered office address and whether it is in the process of being wound up. As a registered company has its own legal personality, the company itself can commit an offence and can therefore be prosecuted.

Partnerships
As a partnership or limited partnership has no legal personality in the context of criminal law, it cannot be prosecuted.[90] In a standard partnership, all of the partners are liable and an information can be laid against one or more of them.

Generally, all the partners should be charged unless there are good reasons for not doing so, e.g. investigation shows that the way in which duties and responsibilities are divided excludes someone from responsibility for an offence, or a partner has done something

---

[85] Under s.20, Health and Safety at Work etc. Act 1974 or the inference drawn from the silence of a person in accordance with the Police and Criminal Evidence Act 1984. This will usually occur at a much earlier stage in the process.

[86] *R. v. Norkett, exp. Geach* (1915) 139 L.T. Jo. 316; *Allen v. Wiseman* [1975] R.T.R. 217. See also s.123, Magistrates' Courts Act 1980.

[87] *Aldi GmbH and Co. KG v. Miulvenna* (1995) J.P. 717.

[88] Registered with the Register of Companies and being a limited company.

[89] Companies Act 1985.

[90] This is not the case in a limited liability partnership which can be prosecuted.

outside the partnership arrangements, such as obstructing an officer in his duties. Separate informations should be laid against each partner in their own names but stating that they are being proceeded against as one of the partners.

Limited partnerships are constructed differently and have two types of partner. The "limited partners" generally do not take part in the management of the business and would not normally be proceeded against unless they actually took part in its management. Proceedings would normally take place against the "general partners" who are liable for the firm's management obligations.

Individuals

Proceedings may be taken against an employee[91] if he contravenes his general duty "to take reasonable care for the health and safety of himself and other persons who may be affected by his acts or omissions at work; and as regards any other duty or requirement imposed on his employer … to co-operate with him so far as is necessary to enable that duty or requirement to be performed or complied with."[92] He must have a contract of employment, written or implied, and any proceedings must be in respect of something he has done whilst at work and carrying out that work in the course of his employment.

If an employer appears to be mainly responsible for the circumstances leading to an offence, then action should only be taken against him. However, if the employer has taken all "reasonably practicable steps" to comply with the law, it may be appropriate to proceed against the employee in the following circumstances:

(a) he failed to comply with safe working practices set out by his employer and which he had been trained to follow;

(b) he obstructed an officer in the performance of his duties;

(c) he had received previous warnings from his employer or the enforcing authority.

---

[91] As defined in s.53, Health and Safety at Work etc. Act 1974.
[92] s.7(a)(b), Health and Safety at Work etc. Act 1974.

Deciding whether to proceed against an employee who has been injured as result of his own failure to comply with health and safety requirements could be less straightforward. For example, if an employee had received adequate training and supervision in a task and then rashly chose to ignore safety rules, resulting in a serious and permanently disabling condition, what value would there be in a prosecution? Quite clearly there would be a *prima facie* case to answer and a conviction would be likely. The issue of whether it was in the public interest to proceed is another matter. If only the individual has suffered, he will have learnt his lesson. If other people might have been affected by his rashness there might be a case for proceeding, although any work colleagues will undoubtedly have learnt something from the events. The details of serious injuries also have a habit of gaining wider publicity through the media and therefore the health and safety message reaches a wider audience. Accordingly, it may not always be necessary to prosecute to achieve the effect that such proceedings can have!

Trades unions recognise the importance of both employers and employees complying with their statutory duties and it is important that there is liaison with trades union representatives where proceedings are being considered against an individual.

Offences by corporate bodies
Section 37 of the Health and Safety at Work etc. Act 1974 states that where an offence under any of the relevant statutory provisions is proved to have been committed by a body corporate with the consent or connivance of, or to have been attributable to any neglect on the part of, any director, manager, secretary or other similar officer of the body corporate, he as well as the body corporate shall be liable to be proceeded against.

The use of the words "consent" and "connivance" indicate that the defendant has relevant knowledge of the facts that constituted the offence and makes a decision based on that knowledge. The responsibility for an offence under this section depends on the nature of the responsibility and authority of the individual in the company's organisation.[93] Accordingly, if proceedings are considered under section 37, as much information as possible should be obtained to demonstrate the duties and responsibilities of the individual. The health and safety policy would be the starting point but contracts of employment, letters advising of changes in responsibility and even the unwritten development of some responsibilities may provide evidence where it can be demonstrated that the individual has actively consented or connived in the neglect. Some of the key issues to consider in deciding if proceedings under this section are appropriate are:

• Did the person have proper control over the matter?

• Were they personally aware of the circumstances involved?

• Were there obvious precautionary steps they could take to prevent the situation?

• Had they been previously warned or made aware of the situation?

• Is more than one person culpable?

Other people
Where an offence is due to the act or default of some other person, then that person may also be charged with an offence, whether or not proceedings are taken against the first person.[94] Such proceedings might include a contractor or a trespasser who had interfered with plant or equipment to such an extent that an accident occurred as a result.

DECISION TO PROSECUTE

Obtaining authority to prosecute

The decision to prosecute needs to be carefully considered in each case and will generally follow the criteria in the enforcing authority's

---

[93] In *R. v. Boal* [1992] 3 All E.R. 177, the manager of a bookshop successfully appealed against a conviction on the basis that he was only responsible for the day-to-day running of the premises. In *Armour v. Skeen* [1977] I.R.L.R. 310, a local authority departmental director, although not a director in the sense of s.37, was prosecuted under s.37 on the grounds that the failure to produce a health and safety policy statement required of him by his authority was due to his neglect.

[94] s.36, Health and Safety at Work etc. Act 1974.

enforcement policy. It will need to be careful in exercising any judgement to proceed outside the terms of that policy. A number of principles have been established that are relevant to the use of local authority prosecution policies:

1. It is proper for enforcement authorities to adopt a policy on the initiation of prosecutions and, having adopted a policy, the policy must be applied, not rigidly, but flexibly and in good faith, taking into account all the circumstances.[95]

2. A prosecutor must not fetter its discretion for the future. It is not permissible, therefore, to take a decision not to prosecute a whole class or category of offender because that would be a fetter on the discretion of the local authority and an abuse of power.[96]

3. Where there is a policy, failure to apply the policy resulting in a decision not to prosecute may, as with an overly rigid adherence to the policy in favour of prosecution, be susceptible to judicial review.[97]

The general power of a local authority to prosecute is contained in the Local Government Act 1972.[98] The decision will usually be taken by a committee, a sub-committee or an officer provided with the delegated power by the council of the authority. In legal proceedings, the decision to take enforcement action may be successfully challenged if the authorisation has not been properly obtained, and it may be necessary to produce that authorisation in court.

In most local authorities, the decision to prosecute will be taken by a service committee given delegated powers by the council. The decision will usually, although not invariably, follow consultation with the authority's solicitor who must be satisfied that the evidence meets the relevant criteria for a prosecution. The details to be sent to the solicitor may include the following:

---

[95] *R. v. Commissioners of Inland Revenue, ex p. Mead* [1993] 1 All E.R. 772.
[96] *R. v. Chief Constable of Devon and Cornwall, ex p. Central Electricity Board* [1982] Q.B. 458.
[97] *R. v. Director of Public Prosecutions, ex p. Chaudhary* [1995] 1 Cr. App. R. 136.
[98] s.222.

- A list of the items submitted.

- A copy of the committee report.

- Details of the elements of the enforcement policy which support the proposed course of action (this is a public document available for the defence to use if there is a departure from that policy).

- A summary of the evidence and the offences involved.

- The lead inspector's statement of evidence, followed by the statements of supporting staff.

- Copies of witness statements, including any from expert witnesses.

- Copies of any photographs, sketches, plans, diagrams or maps, all suitably labelled and containing any relevant measurements and other information necessary to the understanding of the documents.

- Other documentary evidence such as safety policies, accident reports, company letters, memoranda, etc.

- Copies of any relevant advisory letters, notices and warnings.

- Summaries or transcripts of PACE interviews.

- Copies of inspection reports, especially any leading up to the offence.

- Any Companies House search confirming relevant information such as a company name and its registered office.

- Any documents or statements indicating the defendant's previous safety record.

- A list of previous convictions, if any.

- A statement of the authority's costs to date (this will need to be updated at the hearing if costs are to be sought).

Where necessary, two sets of documents should be available, one for the authority solicitor and the other for disclosure to the defence.

It is also important that certain related information is provided, including:

- Information favourable to the defence which may be used to refute the charges.

- Any views of a victim of the offence, e.g. an injured or bereaved person.

- Any aggravating or mitigating factors.

- Any views or representations made by the potential defendant.

The tests

The decision to prosecute is a serious one with potentially significant repercussions for the individuals involved. Accordingly, inspectors proposing to pursue a prosecution should be aware of the tests to be applied before submitting a prosecution report. These are set out in the Code for Crown Prosecutors.[99] There are two stages in the decision to prosecute: the evidential test and the public interest test.

The evidential test

If the case does not pass the evidential test it must not go ahead, no matter how important or serious it may be. The Crown Prosecution Service expects to be satisfied that there is enough evidence to provide a *realistic prospect of conviction*.[100] In practice, this means that the inspector must consider the following:

1. Will a court be satisfied that the evidence is reliable and admissible in court and complies with legal rules, e.g. the use of hearsay evidence and PACE interviews?

2. Is there evidence which might support or detract from the reliability of a confession?

---

[99] Crown Prosecution Service, 2000.
[100] *ibid.*

3. What explanation has the suspect given, does it suggest an innocent explanation and will it stand up in court?

4. Is the background of any witness likely to weaken the prosecution case and does the witness have a suspect motive that might influence their evidence?

5. Are there concerns over the accuracy or reliability of a witness?

A number of these issues might be relevant, especially if someone tries to blame another person to hide their own failings.

A realistic *prospect of conviction* is an objective test, meaning that a jury or bench of magistrates, properly directed in accordance with the law, is likely to convict the defendant of the charge alleged. This is a separate test from the one that the criminal courts must themselves apply. A jury or magistrates' court should only convict if it is sure of a defendant's guilt.

The public interest test
If there is enough evidence to provide a realistic prospect of conviction, then the public interest test can be considered. The factors for and against a prosecution have to be carefully considered and these will probably appear in the authority's enforcement policy statement. The Code for Crown Prosecutors lists some of the public interest factors in favour and against prosecution.[101] Put into a health and safety context, the *factors in favour* might include:

•   A death or serious injury resulted from the contravention.

•   The alleged offence was so serious that significant harm could have occurred.

•   There was a reckless disregard for health and safety requirements.

•   There were repeated or persistent breaches of the law giving rise to significant risk.

---

[101] Crown Prosecution Service, 2000, paras. 6.4 and 6.5.

- Activities were carried on without relevant licences or despite a risk assessment revealing serious risks.

- Significant risks arose from sub-standard management of health and safety requirements.

- There was failure to comply with an improvement or prohibition notice.

- An officer had been intentionally obstructed in the course of his duties.

- Information required had been wilfully withheld, or false information supplied, or there had been an attempt to deceive over a health and safety matter.

The *factors against prosecution* might include:

- The issues were relatively minor and the courts were only likely to impose a nominal penalty.

- The offence was committed as a result of a genuine mistake or misunderstanding and there was minimal risk to health and safety.

- The victim was an employee who had suffered serious injury and little or no benefit would be gained by prosecution. Consideration would have to be given as to whether others had been injured or suffered loss from the individual's actions.

- The defendant was elderly or suffering from significant mental or physical illness, unless the offence was serious or likely to be repeated.

In considering the public interest test, the consequences for the victim of deciding whether or not to prosecute should be taken into account, together with any views expressed by the victim or the victim's family.[102] In any event, victims of an offence, their representatives (including trades union representatives) and their families where necessary should be kept advised of any action being taken, together with its progress.

---

[102] The Code for Crown Prosecutors, para. 6.7.

Formal cautions

If the public interest does not justify a prosecution but there is evidence of a criminal offence, one appropriate course of action is a formal caution.[103] In the HSE's enforcement policy statement, a formal caution is defined as:

> "a statement by an inspector, that is accepted in writing by the duty holder, that the duty holder has committed an offence for which there is a realistic prospect of conviction".

Formal cautions are not the same as those given under the Police and Criminal Evidence Act 1984 prior to asking a suspect questions relating to an alleged offence. Furthermore, a repeat of a breach that has previously been dealt with by a formal caution should normally be the subject of a prosecution. A formal caution should only be considered where a prosecution could be brought, i.e. the case meets the evidential test.

A formal caution is quite often used by local authorities as an alternative to prosecution and appears as part of their enforcement policy statements. It should not be used as an alternative to a weak prosecution case or be confused with a warning letter.

The grounds for deciding to administer a caution would include those public interest factors against prosecution.[104]

The factors to be considered in deciding whether to administer a caution include:

- Is there sufficient evidence available to justify a prosecution? The same standard is required for administering a formal caution as is required for a prosecution.

- The offender must admit their guilt and give their informed consent to the caution.

- If the offender refuses to accept the caution then a prosecution would normally, but not automatically, follow.

---

[103] See "The cautioning of offenders", Home Office circular 18/1994.
[104] See pp.198 and 199 above.

- If the offender has been previously cautioned it would not be usual to administer a further one, especially for the same type of offence, unless the inspector was satisfied that the circumstances were such that a further caution was appropriate.

Although a caution can be administered verbally and recorded in writing, it is usual and preferable to administer it in writing. In the event of any future prosecution, a caution can be produced if the defendant is found guilty. The offender should be advised of this and also be informed that acceptance of a formal caution would be taken into account in any future decision on whether or not to prosecute for any further offences. The caution should be administered adopting a similar approach to an interview,[105] including:

1. Sending an invitation to attend a meeting to explain the process and advising of the implications of accepting a caution.

2. Meeting with the offender and/or their representative (a letter of authorisation may be required).

3. The recipient of the caution must understand the nature of the caution and its consequences, agree to the procedure and admit the offence.

4. Two signed copies of the caution should be made, one for the recipient and one retained by the officer.

Pre-trial matters

The correct drafting of informations, selection of charges, service of summonses and a myriad of related administrative matters are aspects of a prosecution which, although vital to the success of a prosecution, are best left to the lawyers responsible for pursuing the matter through the courts. These detailed aspects of a prosecution case are adequately covered in a number of other documents.[106] As

---

[105] See, generally, pp.174-181.
[106] Notably the *Health and Safety Executive Enforcement Guide (England and Wales)*, updated September 2003. See also *Stone's Justices Manual* published yearly.

this book is intended to provide practical advice to health and safety practitioners, the rest of this chapter seeks to cover the practical elements of the proceedings with which those people will be most involved.

Attendance of witnesses

Cases rest and fall on the attendance and evidence of witnesses, so not only must their evidence be admissible but they must also attend court when required. Some straightforward steps will help to ensure their attendance:

1. Clarify the dates when they will not be available to attend court because of holidays, work commitments, hospital appointments, etc.

2. Obtain their names and addresses and telephone numbers to enable them to be readily contacted.

3. When a court date has been arranged, a letter should be sent to them advising of the date, time and place of the hearing; asking them to confirm their availability and willingness to attend; giving details of how to get to the court; and enclosing details of how to claim expenses.

4. If there is reason to suspect that a witness will not attend the court hearing voluntarily, application should be made to the clerk of the court for a witness summons. The court will need to be given the name and address of the witness and advised of any document the witness is required to produce. It may also be necessary to explain the nature of the witness evidence that is material to the case.[107]

5. The summons can be served by the inspector or another officer, or the police. A witness summons has to be served personally on the witness, not by post.[108]

---

[107] *R. v. Peterborough Magistrates' Court, ex p. Willis and Amos* (1987) 151 J.P. See also *R. v. Marylebone Magistrates' Court, ex p. Gatting and Emburey* [1990] Crim. L.R. 578, where it was held that it was not sufficient that the prospective witness could give evidence of what he saw or heard of an incident leading to the criminal proceedings; *the evidence must be material to the case* of the litigant.

[108] Magistrates' Courts Rules 1981 (S.I. 1981 No. 552), Rule 99(6).

6. If it is intended to call a witness who requires an interpreter, the court should be informed as soon as possible so that it can arrange for a suitable interpreter to attend.

It is possible to use the contents of a witness statement in a trial without the witness having to give evidence. However, certain procedures must be followed for this to take place:

1. Both parties to the proceedings must agree to the use of the statement in this way.

2. The form of the statement must comply with specific requirements.[109]

Although this approach may save time and help the prosecution, it does not of course provide for the witness to be cross-examined. The defence is, therefore, only likely to agree to this process if the information in the statement is straightforward, factual and non-contentious. In most cases the defence is likely to want to give the defendant the best chance of acquittal and insist on the witness being present for cross-examination.

Disclosure of unused material

Although a defendant is entitled to advance information[110] of a prosecution case, local authorities often fail to understand or have proper procedures for the disclosure of unused material. The Criminal Procedure and Investigations Act 1996 regulates the investigation process[111] but also regulates the recording and retention of material found or produced during an investigation. Although the Code of Practice produced under the Act applies to police officers, other investigators must have regard to the provisions of the Code.[112] The Act requires that all reasonable lines of enquiry are pursued in the course of an investigation. This implies the inclusion of lines of enquiry helpful to the defence, and such information has to be disclosed to the defence. There are three disclosure stages:

---

[109] Magistrates' Courts Act 1980, s.5B in the case of committal proceedings; Criminal Justice Act 1967, s.9 in all other cases.
[110] The Magistrates' Courts (Advance Information Rules) 1985 (S.I. 1985 No. 601) govern the service of advance information.
[111] See pp.155-158.
[112] s.26 and para. 1.1 of the Code.

1. *Primary prosecution disclosure* – requiring the prosecution to disclose any unused material which could undermine the prosecution case. This might include:

   (a) contradictory witness statements;

   (b) sampling results which do not appear to support conclusions as to health and safety risks;

   (c) notebook entries which are inconsistent with a witness statement or conclusions.

2. *Defence disclosure* – requiring disclosure of the details of the defence in all trials on indictment. Disclosure is purely voluntary in the magistrates' court. The disclosure is by way of a "defence statement" setting out:

   (a) the terms of the defence;

   (b) those aspects of the prosecution case with which the defendant takes issue;

   (c) the reasons for taking issue.

3. *Secondary prosecution disclosure* – having received the defence statement, the prosecutor must consider whether any of the unused material would assist the defence case. If so, it must make a secondary disclosure of any such material.

The procedure is an ongoing process of review to ensure that additional disclosure is not required.

Cases may occur where the defence has a document, the contents of which have been noted or copied, and the prosecution relies on that evidence as part of its case, e.g. a register or accident record. If the prosecution wishes that document to be produced in court, a recorded delivery letter should be sent to the defendant describing the document and asking for its production in court. For the avoidance of any doubt about the document to be produced, it may be useful to attach a copy to the letter. If the defendant chooses not to produce the document in court, the copy can then be produced. If the defendant is a body corporate, it is preferable to serve a witness summons on the secretary or other person having control of the documents needed, requiring their attendance and production of the documents.

If there are witnesses other than the defendant who have documents the prosecution wishes to be produced, then a witness summons should similarly be served. This may well be the best approach in a case involving several witnesses with documentary evidence helpful to the prosecution, to ensure that all the evidence is produced at the trial.

Evidence in court

The rules of evidence are covered elsewhere but it may be helpful to address some of the practical issues when evidence is presented by witnesses.

Oral and written evidence

There are some factors that should be considered well before an officer sets foot in the witness box so that he is prepared for what is to follow:

• The prosecution will allege in certain breaches of the legislation that the defendant has failed to do all that is reasonably practicable to reduce the risk to health and safety. Equally, the defence will attempt to show that on the balance of probabilities the defendant has done all that is reasonably practicable.

• The defence will usually seek to discredit the witness and/or his evidence.

If the prosecution evidence cannot stand up to these challenges, the lead officer and his witnesses may be in for a rough time. If the preparation for the case has not been thorough from the time that an investigation started through to the court hearing, the defence is likely to pick holes in the evidence. This will be embarrassing and stressful for the witnesses, making it difficult to stay properly focused on giving evidence and the defence lawyer will seize on the opportunity this presents.

There are a number of practical steps that can be taken to minimise these risks:

1. Be thorough in following the rules for obtaining evidence. It is often better to take too many notes than too few to ensure that all relevant material is recorded.

2. It should be assumed that the inspector's notebook may be referred to by him in court if he needs to refer to it to refresh his memory. If he does so, the defence will be entitled to see his notes. If they are not consistent with his statement of evidence, are not contemporaneous, are illegible or are written in his own form of shorthand, then he can expect to be challenged on the accuracy of his notes and his recollection of events. If the notes are not contemporaneous or written up soon after the events to which they refer, the witness may be questioned as to whether they are his own notes or written with the help of someone else. "Contemporaneous" requires facts to be fresh in the memory when written down. This can include notes written several hours after the event, depending on the contents.

3. Rehearse the evidence well before the hearing. There is less chance of the officer's credibility being challenged if he is clearly in command of the facts of the case. It can be advantageous to rehearse the case with the prosecuting lawyer and an experienced senior colleague. As this process allows the other officers to challenge the evidence in the same way as the defence might do, it should reveal any problems, issues, evidential matters and other flaws in the prosecution evidence. There may then be time to address any matters raised or to obtain any further evidence that may be required. If not, at least the lead officer will be prepared for any potentially difficult questions. However, it should be noted that, valuable as this process can be, tailoring the evidence to the case is not allowed!

Giving evidence

The oath

Before any evidence is given, witnesses have, on their oath or affirmation, to promise to tell the truth, the whole truth and nothing but the truth. We all know of cases where a defendant has promised to do this and then proceeded to lie. Prosecution witnesses have been known to do the same. The truth must be told as to do otherwise may amount to perjury. Do not try to win a case that may appear to be going badly for the prosecution by lying, the defence will probably reveal the lie and the court will not be happy about it. The offence of perjury can result in a custodial sentence.

Exhibiting the evidence

Except for statements and depositions, all other documented[113] or real evidence (material objects) needs to be "exhibited" in a statement with each document identified with the initials or name of the person making the statement and the items numbered consecutively. Where possible, the originals should be produced although, subject to certain conditions, the production of a copy will be acceptable.[114]

Where a material object is evidence, it can be produced in court, although verbal evidence can be given where the object is not produced.[115] Where possible, material evidence should be produced as this will add weight to the evidence. Such objects should be produced as exhibits, appropriately marked and sufficiently described in the statement to identify them. If evidence has been destroyed, e.g. as a result of testing, then it is possible to rely on photographs or video recordings. It is important to remember that once an article has become an exhibit, the prosecution has a duty to preserve and retain it for production at the trial[116] and allow the defence reasonable access to it.

Tape and video recordings constitute documents and are therefore subject to the rules regarding admissibility, and they may also be real evidence when produced to show the situation that was actually recorded, e.g. the operation of machinery implicated in an accident or the path of a fork lift truck prior to an accident.[117]

---

[113] "Documents" are defined in the Criminal Justice Act 1988, Sch. 2, para. 5.

[114] For more detail see "Physical evidence in court – exhibiting evidence", HSE Enforcement Guide (England and Wales), updated September 2003.

[115] *Hocking v. Ahlquist* [1943] 2 All E.R. 722.

[116] See *R. v. Lambeth Metropolitan Stipendiary Magistrate, ex p. McComb* [1983] 1 All E.R. 321.

[117] See *R. v. Thomas* [1986] Crim. L.R. 682.

Photographs and sketches can be produced in evidence provided that an acceptable witness can verify their accuracy. Photographs should be individually identified in a statement unless a collection of them is produced, in which case the use of an index to describe them is acceptable. Photographs should be agreed with the defence to ensure that they do not contain prejudicial material. Such evidence, if properly presented, may show the layout of premises, the location of equipment relative to any accident or other offence and may be as useful as a visit to the premises by the court. It is important that any photographs speak for themselves. They should not be subject to subjective comments. A series of photographs showing the scene of an accident or an injury and the relevant circumstances can avoid the need for extensive descriptive prose and can be damaging to the defence. Indeed, it is difficult to object to a photograph produced without comment as it cannot be cross-examined. Similarly, maps and plans showing the detailed location of things, accompanied by a certificate signed by the person drawing them, stating that they are drawn to a specified scale and properly represent the place specified in the certificate, are also acceptable.[118]

Questioning witnesses
The prosecution will obviously have to call all of its evidence before it concludes its case. Only rarely will it be allowed to call evidence afterwards. The evidence should therefore be examined thoroughly with the prosecuting lawyer before the hearing. Nevertheless, if evidence it was intended to introduce has not been brought before the court, the prosecutor must be told well in advance of the close of the prosecution case. It is generally too late afterwards.

When witnesses are called to give evidence, they will be questioned by the lawyer representing the party who has called them. This "examination-in-chief" seeks to put forward all of the information in support of that party's case that the witness knows about. After witnesses have given evidence they may be cross-examined by the other party's lawyer, especially if they have said something which the defence feels they can turn to their advantage.

---

[118] Criminal Justice Act 1988, as variously amended.

A competent defence solicitor will attempt to get a witness to say what suits his case during cross-examination, whilst at the same time trying to avoid letting the witness say things that may harm the defence. The defence will often ask a question, or what turns out to be a question with a number of parts to it, and then ask the witness to answer yes or no. In many cases there is no simple yes or no answer and it is obvious that the defence is trying to avoid allowing the witness to give detailed information which may not be helpful to their case. In such cases the court should be advised that there is no simple yes or no answer, or that there is more than one question to be answered. The witness should then proceed to deal with the question(s) appropriately, telling the whole truth throughout.

The witness will not always be able to recall all details about a case. He should avoid trying to guess about something. If he does, he will not be being truthful and the defence may produce evidence refuting his statement. His credibility will then be damaged and this may affect the outcome of the case. If a witness cannot remember something, he must be completely honest (he has taken an oath to tell the truth!) and say so.

If the cross-examination has not gone well for the prosecution, there remains a further chance to correct any misunderstandings or doubts raised as a result of the answers given to the carefully crafted questions of the defence. After cross-examination, the party calling the witness has the opportunity to re-examine their witness. Questions must, however, be limited to clarifying matters dealt with during cross-examination.

At the end of the cross-examination, especially if things have not gone well for them, the defence lawyer quite often tries to summarise the witness evidence by playing down evidence harmful to the defence case, glossing over or generalising matters of important detail and emphasising those aspects of the evidence favourable to the defence. If what he says is not true or is a distortion of the facts, the witness should say so. The court is unlikely to object as its only interest is justice.

When a health and safety inspector has given evidence and sits in court to hear the remaining evidence, he may find the defence evidence impossible to believe and the comments of the defence solicitor in mitigation a complete farce. He must resist the temptation to try to influence the magistrates or jury by using facial expressions designed to convey his thoughts about the defence evidence. Otherwise he may provide grounds for appeal.

It can be helpful to find out, well in advance of the court hearing, which lawyer will be presenting the case for the defendant. The benefit of this is that the prosecuting lawyer may know about that person, his style of questioning and his general manner in court, as well as any particular difficulties he may have presented to prosecution witnesses in the past. This can help the lead prosecution witness to prepare for the style of questioning, particularly if it is likely to be aggressive.

Use of witness statements and notebooks as a memory aid
At one time it was not possible to use a witness statement in the witness box as a means of refreshing the memory of a witness.[119] It is still best to prepare sufficiently beforehand, if practicable, and avoid the suggestion of a failing memory when questioned by the defence. However, a witness may, subject to the court's approval, look at a statement made earlier if it accepts that:

(a) the witness cannot recall the details of events because of the lapse of time;

(b) he made a statement nearer the time when he could remember the events described;

(c) he had not read the statement before coming into the witness box;

(d) he wishes to look at the statement before he continues to give evidence.

A witness may refresh his memory by referring to a contemporaneous document such as his notebook. The defence is entitled to see documents from which he has refreshed his memory

---

[119] This changed following the case of *R. v. da Silva* [1990] 1 W.L.R. 31.

and the witness can be cross-examined on the contents. It is prudent to ensure, when making contemporaneous notes, that they are sufficiently detailed and written in clearly legible handwriting to avoid awkward questions on the accuracy and interpretation of them in court. The contents must have been written at the time of the events or so shortly afterwards that the facts are still fresh in the witness's memory.

Use of expert witnesses[120]
In some cases expert evidence is necessary to the success of a case, e.g. the taking and analysis of samples.[121] It cannot, however, replace the need for factual evidence and legal argument.

The standards expected of an expert by a court and, therefore, by any local authority seeking to use the evidence of an expert in court proceedings include:

- Expert evidence should be, and should be seen to be, the independent product of the expert, uninfluenced by the exigencies of the litigation in which he is involved.[122]

- An expert should provide independent assistance to the court in an unbiased way. He should never act as an advocate for the party who has called him.[123]

- An expert should state the facts or assumptions upon which his opinion is based. He should consider material facts which would detract from his considered opinion.[124]

- An expert should make clear when a matter falls outside his expertise.

- If an expert's opinion is not properly researched because of an insufficiency of data or other reason, he must say so.[125]

---

[120] For general advice in appointing expert witnesses, see *Successful Use of Expert Witnesses*, Burn, S., Shaw & Sons Ltd.

[121] Valuable advice on the use of experts is contained in the *HSE Enforcement Guide (England and Wales)*, updated September 2003, HSE.

[122] *Whitehouse v. Jordan* [1981] 1 W.L.R. 246.

[123] *Pollvite v. Commercial Union Assurance Co. plc* [1987] 1 Lloyds Rep. 379; *Re J.* [1991] F.C.R. 193.

[124] *ibid.*

[125] *Derby and Co. Ltd. v. Weldon and Others, The Times,* November 9, 1990.

These matters are important when selecting an independent expert to assist in the collection of evidence, the presentation of that evidence or to give an expert opinion. Care should be taken to appoint those who have a demonstrable track record in dealing with the issues at stake and can produce evidence of their ability to present their evidence effectively in court. If both parties to a case use expert evidence, the case may hinge on who gives the most believable evidence in court. If experts in an investigation are producing factual evidence, e.g. examination of machinery or sampling and analysing materials or substances, their evidence may be similar. Where expert evidence may differ is in the interpretation of results or their opinion as to the impact of the evidence presented. If they are merely giving opinion evidence, the expert must hear what witnesses say. They are therefore entitled to sit in the court to hear the factual evidence of the witnesses. In their evidence they can then comment on the witnesses' evidence, e.g. by giving a view on the most likely risk to health and safety arising from a series of events.

The court will decide if evidence can be given by someone as an expert witness. That person will have to give evidence as to his expertise, e.g. qualifications, experience, published technical papers. A general expertise is not sufficient, and a health and safety inspector would only be regarded as an expert witness if he has specialised in a particular sphere of activity and reached a high level of expertise. That expertise must also be up-to-date. Beware of the defence advocate who invites an inspector to say he is an expert in a particular field because of his qualifications and then seeks to tear his evidence to shreds in cross-examination and cast doubt on his credibility as a witness.

If using an external expert witness, it is usually best to encourage him to give his evidence in layman's language so that it can be understood by all parties, not just by the other side's expert.

Experts' reports can be given in evidence even if the expert does not give oral evidence.[126] The permission of the court is required and, in making its decision, it will have regard to the contents of the

---

[126] Criminal Justice Act 1988, s.30.

report, the reasons for not giving oral evidence, the possible unfairness to the defendant, whether it will be possible to challenge the report and any other relevant matters.[127]

Defendants who remain silent

Defendants who exercise their right to silence during interview may also do so in court. In determining whether the accused is guilty of the offence, the court is entitled to draw whatever inferences as appear proper from the failure to give evidence or a refusal to answer questions without good cause.[128]

SENTENCING

The Court of Appeal has made recommendations for the presentation of health and safety prosecutions to help the courts to sentence on a correct factual basis. It particularly recommended that when proceedings are commenced, the prosecution should not merely list the facts of the case, but also a schedule of aggravating and mitigating factors to assist in sentencing.[129] The court has also provided guidance on assessing the gravity of the breach of the law in each case. It said:[130]

> "In assessing the gravity of the breach it is often helpful to look at how far short of the appropriate standard the defendant fell in failing to meet the reasonably practicable test."

Accordingly, when listing the aggravating and mitigating factors in a case,[131] inspectors should take this judgement into account. The aggravating and mitigating factors to be taken into account in health and safety cases may include the following,[132] but the actual factors will vary from case to case.

---

[127] Criminal Justice Act 1988, s.30(2).
[128] Although the court cannot convict the defendant solely on the basis of an inference drawn from the failure to give evidence.
[129] The judgement arose from the case of *R. v. Friskies Petcare U.K. Ltd.*, [2000] 2 Cr. App. R (s) 401 and the aggravating features as set out in the case of *R. v. F. Howe and Son (Engineers) Ltd.* [1999] 2 All E.R. 249.
[130] In the *Howe* case.
[131] Known as the "Friskies Schedule".
[132] Taken from the *Howe* judgement and also referred to in Operational Circular OC 178/2, HSE Field Operations Directorate. For an example of a list of aggravating and mitigating circumstances, see Health and Safety Executive Enforcement Guide (England and Wales), *Sentencing and costs – model examples.*

Aggravating factors

- The standard of care is the same regardless of the size of the defendant organisation.

- The degree of risk and the extent of danger are relevant factors.

- Although it is often a matter of chance whether death or serious injury results from a serious breach, where death does occur it is an aggravating factor and the sentence should reflect public disquiet at the unnecessary loss of life.

- Deliberately running a risk/cutting corners in order to save money/increasing profit is a seriously aggravating factor.

- Failure to heed warnings is an aggravating factor.

- Breaches continuing over a period of time is an aggravating factor.

Mitigating factors

- Prompt admission of guilt and a timely plea of guilty.

- Steps taken to remedy deficiencies after they are drawn to the defendant's attention.

- A good safety record.

Penalties

Fines and imprisonment

The current penalties in relation to offences under the Health and Safety at Work etc. Act 1974 and its associated regulations are as follows:

- Failing to discharge a duty under sections 2-6 carries a maximum fine on conviction in the magistrates' court of £20,000,[133] and in the Crown Court the maximum penalty is an unlimited fine.

- Breaches of the terms of an improvement or prohibition

---

[133] s.33(1A).

notice, or of a remedial order[134] made by the court, carry a maximum fine on conviction in the magistrates' court of £20,000 and/or six months' imprisonment.[135] On conviction in the Crown Court the maximum penalty is an unlimited fine and/or two years' imprisonment.

- For offences triable only in the magistrates' court (inquiries and the powers of inspectors)[136] the maximum fine is level 5 on the standard scale.[137]

- Breaches of sections 7-9 attract a maximum penalty on conviction in the magistrates' court of £5,000 and on conviction in the Crown Court an unlimited fine.[138]

- Breaches of health and safety regulations carry a maximum fine on conviction in the magistrates' court of £5,000 and on conviction in the Crown Court an unlimited fine.

Compensation orders

The magistrates' and Crown Courts have the discretion to make an order requiring a convicted defendant to pay compensation in respect of any personal injury, loss or damage resulting from the offence.[139] If it decides not to do so, it must give reasons for its decision.[140]

The maximum sum that magistrates can award is £5,000 for each offence but the Crown Court can impose an unlimited sum. A compensation order can be imposed in addition to any separate sentence or as a penalty in its own right.[141] If a penalty and a compensation order are imposed and the offender does not have the means to pay both, the compensation order takes priority.[142] The amount of compensation has to be what the court considers appropriate and the court must have regard to the defendant's

---

[134] s.42.
[135] s.33(2A).
[136] s.33(1)(d), (e), (f), (h) and (n).
[137] Currently £5,000.
[138] s.33(3).
[139] Powers of Criminal Courts (Sentencing) Act 2000, s.130(1).
[140] *ibid.*, s.130(4).
[141] *ibid.*, s.130(11).
[142] *ibid.*, s.130(12).

means.[143] Both the prosecution and the defence can make representations concerning the loss suffered by the victim.

There is nothing to stop the prosecution suggesting a compensation order where these are available. The prosecution could assist the court by providing any evidence of injuries, loss or damage incurred.

Disqualification of directors
In some cases it may be appropriate to invite the court to make a disqualification order against a director convicted of indictable offences.[144] This is quite an extreme step but it may be appropriate in cases where a director has consistently failed to take steps to comply with the law, thereby exposing his employees to repeated and prolonged risks to their health and safety.

Remedial action
Where a person is convicted of an offence under any of the relevant statutory provisions in respect of matters which appear to the court to be in the power of the offender to remedy, the court may, in addition to or instead of imposing any punishment, order him, within a specified time, to take steps[145] to remedy the matters.

The time fixed for the order may be extended[146] and a person ordered to remedy matters shall not be liable under any of the relevant statutory provisions in respect of those matters in so far as they continue during the time fixed by the order.[147]

---

[143] The court can ask the defendant to provide appropriate details: see *R. v. Phillips* (1998) 10 App. R (S) 419 and the Powers of Criminal Courts (Sentencing) Act 2000, s.130(11).
[144] Company Directors Disqualification Act 1986, s.2(1).
[145] Health and Safety at Work etc. Act 1974, s.42(1).
[146] *ibid.,* s.42(2).
[147] Or any time extension, s.42(3).

Chapter 5

ACCIDENTS

WHAT IS AN ACCIDENT?

There are many definitions of what constitutes an accident. Most seem to refer to them as being unforeseen, unintended or unplanned events resulting in personal injury, damage to property, loss of production or a combination of some or all of these. In a health and safety at work context and having regard to the principal causes of accidents, the author prefers to consider an accident as being an unintended event resulting in ill-health/personal injury or the likelihood of it, which arises from management or human error because someone failed to consider the consequences of their action or inaction.

ACCIDENT STATISTICS AND TRENDS

Information on the nature and causes of reported accidents, together with a comparison of the injury risk with inspection activity, can be found in the HELA report *National Picture: Health and safety in local authority enforced sectors.*[1] It may be useful to divide some of the principal findings in this report into the main accident reporting categories contained in the Reporting of Injuries, Diseases and Dangerous Occurrences Regulations 1995.[2]

Fatal injuries (employees)

- Fatal injuries to employees in the mainly local authority enforced sectors have increased from 9 to 14 when comparing 2002/03 with the previous year.

- Over the past 10 years, the number of fatal injuries has fluctuated, with no overall trend.

- Fatal injuries to employees have increased in office-based industries and wholesale premises, when comparing 2002/03 provisional figures with 2001/02.

---

[1] October 2003, HSE.
[2] S.I. 1995 No. 3163. Some of the figures for 2002/2003 are provisional.

- The number of fatalities has decreased in retail, residential care homes and leisure activities.

- There is an increasing trend in fatal injury rates in the four years to 2002/03.

Fatal injuries (members of the public)

- In 2002/03, provisional figures indicate 30 fatal injuries to members of the public, the same as the previous year.

- Since 1996/97 there has been an upward trend in the number of fatalities to members of the public. The large increases in fatalities in 2001/02 and 2002/03 are primarily due to the increase in fatalities of residents in residential care homes.[3]

- Fatal injuries in the retail sector have increased from one in 1999/2000, to four in 2002/03.

- In hotels, the number of fatalities has decreased steadily. In the remaining industries within the mainly local authority enforced sector, figures have fluctuated in recent years with no overall trend.

Non-fatal injuries (employees)

- The number of reported major injuries to employees in 2002/03 provisionally increased to 6,629, a 9% rise on the previous year.

- Reported over-3-day injuries also increased to 28,148, an 8% increase on the previous year.

- Both reported major and over-3-day injuries were at their highest since 1996/97 when the Reporting of Injuries, Diseases and Dangerous Occurrences Regulations 1995 were introduced.

- Reported non-fatal injuries to employees in residential care homes have increased steadily since 1996/97, from 602 to 1,296 in 2002/03.

---

[3] These increases will no doubt be influenced in part by an increase in the number of care homes and the general frailty and poor state of health in their elderly residents.

- Reported non-fatal injuries to the self-employed decreased slightly to 155 in 2002/03 compared with the previous year. However, the reporting level for the self-employed was only 5% in 2001/02.[4]

Non-fatal injuries (members of the public)

- Provisional reported non-fatal injuries to members of the public for 2002/03 decreased to 4,010, a fall of 9% compared with the previous year and a fall of 47% since 1999/2000.

- However, there was an increase of 68% in injuries to members of the public in residential care homes in 2002/03, although the numbers fluctuate yearly.

Work-related ill health

It was also estimated that 437,000 individuals in Great Britain whose current or most recent job in the previous eight years was in a mainly local authority enforced sector were suffering from an illness which they believed was caused or made worse by their job.[5]

Riskiest industries and occupations for injury and ill-health

In considering the planning of local authority enforcement activity using scarce resources, it may be useful to consider what the *National Picture* report[6] considers to be among the riskiest industries and occupations, although this information will undoubtedly have to be supplemented by local knowledge.

The highest rates of reportable non-fatal injuries occur in the social work with accommodation sector, e.g. care homes, recreation, culture and sporting activities, and hotels and restaurants. The lowest rates occur in the wholesale and retail (excluding motor vehicles) sectors.

In the mainly local authority enforced sector, the riskiest occupations for injury and ill health were found to be transport drivers; personal

---

[4]   These figures are unlikely to be anything like an accurate representation of the true accident figures for this sector of employment.

[5]   *National Picture 2003: Health and safety in local authority enforced sectors,* p. 6, HSE, October 2003.

[6]   *ibid.,* pp.18 and 19.

service occupations; skilled trade occupations; process, plant and machine operatives; and senior managerial jobs.[7]

There are two main sources of information on workplace injury, reports provided under the Reporting of Injuries, Diseases and Dangerous Occurrences Regulations 1995 and the Labour Force Survey developed by the Health and Safety Executive to complement the regulations.

The Executive has combined local authority inspection data and the injury statistics to form a view as to whether local authority activity is being targeted to those employment sectors where the risks of injury are high. It has formed several conclusions, some of which recognise the pressures on the use of local authority resources:[8]

1. There is an overall downward trend in the rates of inspection in all premises.

2. Catering premises continue to attract the highest rate of visiting, reflecting visits made in conjunction with food hygiene regulations. However, the relative priority of visits to catering premises has decreased over the past five years.

3. Local authorities have generally maintained the relative priority of visits to residential accommodation and wholesale premises over the last five years. There is a high level of fatalities in these premises, either as the number of fatal injuries to members of the public or rate of fatal injury to employees. There is also a high rate of non-fatal injury to employees in residential accommodation.

4. Local authorities have also maintained the relative priority of visits in consumer/leisure and retail premises, within a falling rate of visiting. The level of fatal injuries is relatively high in consumer/leisure premises, either as a rate of injuries to employees or as the number of deaths to the public. In retail premises, the number of non-fatal injuries to the public is high.

[7] *National Picture 2003: Health and safety in local authority enforced sectors,* Tables 13 and 23, October 2003, HSE.
[8] *ibid.,* p. 24.

5. Office premises have the lowest rate of non-fatal injuries to employees and the public, and also the lowest inspection rates.

Musculoskeletal disorders (bone, joint or muscle problems) were the most commonly reported illnesses ascribed to the current or most recent job in the local authority enforced sector in the preceding eight years, affecting an estimated 182,000 people. Stress, depression or anxiety followed, with an estimated 154,000 people affected.[9]

An estimated 6.1 million working days (full day equivalent) were lost through illness ascribed to a current or most recent job in 2001/02 and 2,124,000 days were lost due to workplace injury, all in the local authority enforced sector.[10]

PRINCIPAL TYPES AND CAUSES OF ACCIDENTS

Types
The most commonly reported types of accident in the local authority enforced sector are slips, trips and falls, and transport related incidents.

Causes
The causes of accidents are many and varied and often result from a combination of factors. The causes may be direct, e.g. machinery not being provided with suitable guards, or indirect, e.g. untrained employees or lack of proper instructions. The following reflect some of those causes but there may well be others:

Physical and mental ability
Some tasks require physical strength and dexterity and/or the mental agility to cope with complex tasks requiring an alertness of mind. The risk of accidents when performing such tasks is increased if those employed on them do not have those particular skills, or those attributes are weakened in some way, e.g. the repetitiveness of the task or illness of the individual.

---

[9] *National Picture 2003: Health and safety in local authority enforced sectors,* p. 2 and Table 17.
[10] *ibid.,* p. 2.

Attitude of employer/employee
The lack of safe procedures, acceptance of careless behaviour and willingness to accept standards of care and performance below the norm for the type of work involved.

Supervision
Inadequate supervision, the allocation of supervisory activities to those without sufficient practical and personal skills, and the lack of clearly defined responsibilities.

Training
Inadequate training for the tasks and responsibilities involved, including the use of untrained staff, agency, seasonal and part-time workers.

Operational guidance and instruction
A regular problem is the lack of adequate operational instructions for plant and equipment; inadequate or unclear instructions and rules where these are available; insufficient briefing of staff, especially when changes in procedures are introduced; insufficient checks by supervisory staff on the adherence to rules and instructions; failure to review practices and procedures regularly; lack of refresher training; and inadequate disciplinary procedures when things go wrong. These can be particular issues when external contractors are engaged on unfamiliar work areas.

Working area design
Poor organisation of work flow, congestion and badly designed working areas can lead to insufficient space for safe working. Inadequate storage, e.g. excessive height of stored materials and difficulty of gaining access to regularly required materials, may lead to falls from height and the use of unsafe means of access to storage areas. Congested passages and gangways may lead to trips, falls and transport accidents.

Methods of operation
The lack of clear operational systems of work, often resulting from failure to analyse activities properly to produce the safest and most efficient method of operation.

Maintenance arrangements

All work and safety equipment must be kept in good repair and efficient working order. The absence of properly planned systems of inspection, maintenance and repair, together with a failure to understand the risks associated with such failings, has resulted in numerous serious and fatal accidents.

Environmental conditions

Good air quality free of harmful contaminants, adequate heating, lighting, ventilation and humidity are essential to a healthy working environment. High levels of noise and vibration may cause stress, poor concentration and difficulty in hearing the spoken word, especially instructions and warnings, leading to a lack of awareness of safety problems and accident risks.

Building structures

The structural features of a building may affect the likelihood of accidents, e.g. sloping floors, especially if wet, greasy or damaged may give rise to a greater risk of slips and falls; poorly laid out working areas may have dark or concealed areas; staircases and higher floors or platforms, where not adequately guarded, may give rise to an increased risk of falls by persons or equipment.

Safety of machinery

The inherent dangers of some machinery; poor guarding and safety mechanisms; the lack of maintenance leading to contact with moving parts; and insufficient inspection and maintenance often results in accidents.

Cleanliness

The absence of well-maintained cleaning schedules and related supervisory arrangements will result in dirty working conditions and an increased risk of infection, slips and falls on floors, and machinery that breaks down more often. In such cases, where the employer appears uninterested in the working conditions of his employees, there is likely to be an increased tendency for the employees themselves to take less care for their own safety and welfare.

Personal protective equipment

Accidents occur as a result of inadequate training in the use of personal protective equipment; the provision of equipment unsuitable for the purpose or insufficient to meet the specific needs of individuals; poorly designed equipment; and inadequate supervision.

Risk assessments

The lack of risk assessments means that employees often have no real idea what hazards they face, let alone what preventative steps they should take to protect themselves from harm.

Stress

There are increasing numbers of work-related stress incidents, some of which may lead to accidents, others resulting in absence due to stress. Much of the stress has been laid at the door of organisations who appear unwilling to recognise it as a consequence of unsatisfactory working conditions.

Some typical incidents

It may be useful to refer briefly to some reported accidents and the employer failings to demonstrate some of the above-mentioned issues. The following are selected as the work activities are commonly encountered in premises subject to local authority enforcement:

1. An employee suffered a crushing accident resulting in a badly cut hand. *No risk assessments* had been conducted, *no manual handling assessments* had been carried out and it was *unclear who had responsibility for health and safety.*[11]

2. A worker was killed when he fell from unsafe scaffolding. The *method statement for the scaffolding work was insufficient*, there was *no safe system of work* and *inadequate supervision.*[12]

3. An employee was crushed to death when 14 desktops fell on to him while they were being unloaded from a lorry. There had

---

[11] *Environmental Health News*, May 16, 2003, p. 7
[12] *ibid.*

been *no risk assessment* and *no safety procedures* in place at all.[13]

4. A young child died in the care of a day nursery after being fed food to which he was allergic. There was *insufficient staff training* and *no proper risk assessment* had been conducted.[14]

5. An employee lost his hand when it was dragged into a mincing machine. Equipment was found to be *dangerous and badly maintained*, there was *no machinery guard*, *no risk assessment* and *no safe system of work*.[15]

6. A young girl was crushed to death by a collapsing carport at a garden centre. There was a *lack of health and safety training*, *safety devices were deliberately bypassed* and *equipment was in severe disrepair*.[16]

7. Two employees suffered injuries associated with the operation of vehicle tail lifts. There had been *no information, instructions or training*, the activities had *not been subject to a risk assessment* and there was no safe *system of work*.[17]

8. Several supermarket employees complained of repetitive strain injury. There was a *failure to train staff adequately*, a *risk assessment was not acted on* and relevant *safety monitoring was not carried out*.[18]

9. A warehouse employee was struck and injured by a forklift truck. The driver was not *trained or authorised* to drive the truck.[19]

10. An employee suffered nerve damage and severe bruising after falling down an unguarded trapdoor. There was *no safe system of work* and *no risk assessment* had been conducted.[20]

---

[13] *Environmental Health News*, December 13, 2002, p. 7.
[14] *Environmental Health News,* December 5, 2003, p. 3.
[15] *ibid.,* p. 8.
[16] *Environmental Health News*, November 14, 2003, p. 7.
[17] *Environmental Health News*, July 11, 2003, p. 7.
[18] *ibid.*
[19] *ibid.*
[20] *ibid.*

11. A young person was crushed by a reversing articulated lorry at a warehouse. There was found to be an *unsafe system of work* and *no young person's risk assessment* had been carried out.[21]

12. An animal keeper was crushed to death at a zoo. There was *no documented system of work, no up-to-date risk assessment* and *no written health and safety policy.*[22]

13. An employee sustained serious leg injuries from an unguarded cutting machine. The employer *had failed to carry out a risk assessment.*[23]

14. An elderly resident was asphyxiated when her neck became trapped in rails at the side of her bed. There was *a failure to make a suitable and sufficient risk assessment.*[24]

15. An employee was severely scalded by steam escaping from an oven door. The employer had *failed to assess the foreseeable risk, carry out a proper risk assessment or provide adequate training.*

In looking at a comprehensive range of reported accidents, it is quite clear that the majority of reported accidents at work which result in prosecution involve failures to:

•   Carry out proper risk assessments.

•   Implement adequate systems of work.

•   Provide health and safety policies.

In many cases the accidents referred to occur on the premises of large national companies, or local employers who have traded from the same premises for many years. This raises serious questions about the health and safety management of the national companies. There are also a significant number of cases where previous accidents have not been reported over a considerable

---

[21] HELA Annual Report 2002, p. 72, HSE.
[22] HELA Annual Report 2002, p. 72, HSE.
[23] *ibid.,* p. 74.
[24] *Environmental Health News,* January 31, 2003, p. 9.

period of time and/or accidents have occurred where employers have repeatedly failed to comply with the enforcing authority's requirements or well-established Codes of Practice. This may raise questions about whether the enforcing authorities have failed to identify, by regular inspection, serious failings which appear to have continued for some considerable time, and why positive enforcement action has not been taken in cases where it is known that an employer is in default of legal requirements. Such failures expose authorities to the risk of charges of failing to discharge their statutory responsibilities, and possible intervention by the Health and Safety Executive.

REPORTING OF INJURIES, DISEASES AND DANGEROUS OCCURRENCES REGULATIONS 1995[25]

These regulations place duties on employers with regard to the notification and reporting of specified injuries, diseases and particular dangerous occurrences. They apply to accidents, etc. occurring in both the local authority and Health and Safety Executive enforced employment sectors. The specific aspects of the regulations dealt with solely by the Executive are not covered here.

Notification and reporting of injuries and dangerous occurrences

Regulation 3(1) provides that, subject to regulation 10,[26] where:

"(a) any person[27] dies as a result of an accident[28] arising out of or in connection with work;

(b) any person at work suffers a major injury[29] as a result of an accident arising out of or in connection with work;

(c) any person not at work suffers an injury as a result of an accident arising out of or in connection with work and

---

[25] S.I. 1995 No. 3163. At the time of writing the regulations were being reviewed.
[26] Certain restrictions on the application of reg. 3.
[27] "Any person" – the duty to notify does not just relate to employees the self-employed, etc. but extends to all visitors, e.g. shop customers: *Woking Borough Council v. BHS plc* (1994) 93 L.G.R. 396; (1994) 159 JP 427, QBD. This definition will also cover by-standers or passers-by.
[28] Defined in reg. 2. The definition is very limited.
[29] Defined in reg. 2 and Sch. 1.

that person is taken from the site of the accident to a hospital for treatment in respect of that injury;

(d) any person not at work[30] suffers a major injury as a result of an accident arising out of or in connection with work at a hospital; or

(e) there is a dangerous occurrence,[31]

the responsible person[32] shall –

(i) forthwith notify the relevant enforcing authority[33] by the quickest practicable means;[34] and

(ii) within 10 days send a report thereof to the relevant enforcing authority on a form approved for the purpose ... unless within that period he makes a report to the Executive by some other approved means."

Regulation 3(2) specifies that[35] "where a person at work is incapacitated for work of a kind which he might reasonably be expected to do, either under his contract of employment, or, if there is no such contract, in the normal course of his work, for more than three consecutive days (excluding the day of the accident but including any days which would not have been working days) because of an injury resulting from an accident arising out of or in connection with work (other than one reportable under paragraph (1)), the responsible person shall as soon as practicable and, in any event, within 10 days of the accident send a report thereof to the relevant enforcing authority on a form approved for the purposes of this regulation, unless within that period he makes a report thereof to the Executive by some other means so approved."

Regulation 3(2) only applies to injuries resulting from accidents to people who are at work and to injuries not reportable under

---

[30] e.g. a member of the public.
[31] Defined in reg. 2 and Sch. 2.
[32] Defined in reg. 2.
[33] As defined in the Health and Safety (Enforcing Authority) Regulations 1998, S.I. 1998 No. 494.
[34] e.g. by telephone or to the HSE's Incident Contact Centre at Caerphilly, Wales.
[35] Subject to the conditions of reg. 10.

regulation 3(1). These are often, but not always, *relatively* minor injuries which result in the injured person being away from work *or* unable to perform the full range of their normal duties for more than three days.[36]

Regulation 2(2) states that "any reference to an accident or a dangerous occurrence which arises out of or in connection with work includes a reference to an accident, or as the case may be, a dangerous occurrence attributable to the manner of conducting an undertaking, the plant or substances used for the purposes of an undertaking and the condition of the premises so used or any part of them."

The definition of an accident
The definition in regulation 2 includes "an act of non-consensual violence done to a person at work". This includes acts of violence against an employee, a matter of increasing concern for employees, employers and trades unions. Accordingly, a major injury suffered by a bank employee as a result of an assault during a robbery or a health and safety inspector attacked by a suspect he was interviewing would be regarded as an accident and must be reported.

Only physical injuries arising from acts of violence suffered by people at work are included in the definition of accident.[37] Some examples to clarify the situation might include:

- Violence and a resultant injury inflicted on a member of the public by an employee *would not be regarded as a reportable accident*. It could of course result in the employee being prosecuted in the criminal and civil courts!

- An employee suffering from shock following a physical assault on him, leaving him unable to carry out his normal range of duties for over three days, *would be reportable.*

---

[36] For advice on calculating when "more than three days" have elapsed, see *A Guide to the Reporting of Injuries, Diseases and Dangerous Occurrences Regulations 1995*, pp. 16 and 17, HSE, 1999.

[37] See the guide to RIDDOR, para. 13. This would not apply to some types of professional sport injuries where taking part implies the acceptance of a degree of violence and risk of personal injury.

- A case of an employee suffering from shock after witnessing a violent assault on a health and safety inspector *would not be reportable*, although the assault on the inspector might be reportable if the inspector was at work.

The phrase "arising out of or in connection with work"[38]
This phrase is used in regulation 3 and is expanded in regulation 2 to explain its meaning and its scope. It is important for enforcing inspectors to understand this, especially in deciding which accidents to investigate and also when advising employers on their responsibilities for accident reporting (it is well recognised that under-reporting is prevalent and some authorities direct resources into publicity on the statutory requirements). The phrase is also relevant to local authorities as employers.

There are three elements to the definition:[39]

1. *The manner of conducting the undertaking.* This includes the way in which work is being carried out, its organisation, supervision and work practices, e.g. a congested passageway may give rise to a fall. It is possible that a health and safety inspector who fails to take elementary safety precautions or ask to see a risk assessment for premises he is inspecting could expose himself to the risk of an accident.

2. *The plant or substances used for the purpose of the undertaking.* This might include machinery or equipment; lifts; ventilation plant; or chemical substances, e.g. where someone suffers injury from an unguarded machine, or a health and safety inspector fails to use chemical sampling equipment correctly, exposing himself to the risk of inhaling toxic fumes.

3. *The condition of the premises used by the undertaking or of any part of them.* This includes the structure or fabric of the premises, its condition and design, e.g. a building under repair or renovation may be structurally unsafe with a risk of material falling on employees, passers-by or an unwary enforcement officer.

---

[38] reg. 2(2)(c).
[39] *ibid.* See also pp. 11 and 12 of the RIDDOR guide.

From an inspector's viewpoint it is always useful to ascertain whether a risk assessment has been conducted on premises he is inspecting, before he starts the inspection. He should not, however, rely on such evidence as a definitive indication of risks. The assessment may have been poorly conducted. Whilst he may wish to refer to it, he should rely on his judgement and experience and asking probing questions.

Reporting the death of an employee

Regulation 4 requires[40] that where an employee, as a result of an accident at work, has suffered an injury reportable under regulation 3 which is a cause of his death within one year of the date of that accident, the employer must inform the relevant enforcing authority, in writing, of the death as soon as it comes to his knowledge, *whether or not the accident has been reported under regulation 3.*

As this regulation only applies to employees, the death is not reportable if someone else dies from an injury reportable under regulation 3, e.g. a visitor or passer-by.

Reporting of cases of disease

Regulation 5(1) requires[41] that, where a person at work suffers from any one of a number of specified occupational diseases and his work involves one of the related specified activities,[42] the responsible person[43] must forthwith send a report to the relevant enforcing authority on an approved form[44] unless he forthwith makes a report to the Executive by some other approved means.

The requirement for a responsible person to report a case of a reportable disease only applies if:

(a) they receive, in respect of an employee, a written diagnosis of the specified disease from a registered medical practitioner *and* the employee's job involves the specified work activity; or

---

[40] Subject to reg. 10.
[41] Subject to paras (2) and (3) and reg. 10.
[42] Part 1 of Sch. 3.
[43] Defined in reg. 2, generally the employer or other person having control of the premises at which the event occurred.
[44] F2508A available from HMSO.

(b) in the case of someone who is self-employed, that person has
     been informed by a medical practitioner that they are suffering
     from one of the specified diseases.

In the latter case, the self-employed do not usually need written
statements when they are off work through illness and it is
sufficient for them or another person to report the matter to the
relevant enforcing authority.

Reporting of gas incidents

This regulation applies mainly to activities controlled by the
Health and Safety Executive but may also apply to certain premises
under local authority control.

Regulation 6(1) requires that whenever a conveyor of flammable
gas through a fixed pipe distribution system, or a filler, or supplier
(other than by means of retail trade)[45] of a refillable container
containing liquefied petroleum gas receives notification of any
death or major injury which has arisen out of any of these
processes, that person must notify the Executive forthwith, and
send a report to the Executive on an approved form within 14 days.

Nothing under the regulation is reportable if it is notifiable or
reportable elsewhere in the regulations.[46]

Records

Regulation 7(1) requires the responsible person[47] to keep a record
of any event notifiable under regulation 3,[48] cases of disease
reportable under regulation 5(1)[49] and any other particulars
approved by the Executive to show that the reporting requirements
have been complied with.[50]

If any record of deaths, injuries at work or disease are kept for any

---

45 This presumably applies the regulation to the wholesale trade which may include
    local authority controlled premises.
46 reg. 6(3)(a).
47 Defined in reg. 2. HMSO publishes an Accident Book meeting this requirement.
48 See pp.227 and 228, reg. 7(1)(a).
49 reg. 7(1)(b).
50 reg. 7(1)(c). Details of the particulars to be kept in the records are contained in Sch.
    4 to the regulations.

other purpose, then provided it covers the injuries and particulars required by the regulations, it will be sufficient for the purpose of the regulations.[51] The kind of alternative records might include the following:

- Copies of completed forms relating to accidents and diseases might be kept in a separate file to any record book and contain more detailed information for examination and analysis.

- The information could be stored on computer provided it was readily accessible for examination and copying.

- The accident record required for social security purposes,[52] provided accidents reportable under RIDDOR are identified, together with the required details.

The required records must be kept either at the place of work to which they relate or at the usual place of business of the responsible person. The entries must be kept for at least three years from the date on which they were made[53] and, if required, extracts must be sent to the enforcing authority on request.[54] Such information may be required under the powers in section 20 of the Health and Safety at Work etc. Act 1974.

Keeping records of this kind should enable employers to analyse their performance in controlling incidents reportable under the regulations. It can assist in carrying out risk assessments and introducing risk control measures. An employer's response to reportable incidents is also of value to enforcing authorities in deciding whether an employer is meeting his responsibilities for protecting the health and safety of his employees.

Defences

It is a defence in proceedings against any person for an offence under the regulations for that person to prove that *he was not aware* of the event requiring the notification or report and that he had

---

[51] reg. 7(2).
[52] Under the Social Security (Claims and Payments) Regulations 1979, S.I. 1979 No. 628.
[53] reg. 7(3).
[54] reg. 7(4).

taken *all reasonable steps* to have all such events brought to his notice. Presumably, to use this defence successfully it would be necessary to show that written instructions had been given, staff had been trained in what was required and checks were conducted to ensure those instructions had been followed.

Exemptions

The Executive can, subject to conditions, grant exemption from any requirements of the regulations provided it is satisfied that the health and safety of those affected by the exemption will not be prejudiced in consequence of it.[55]

The Incident Contact Centre

The Incident Contact Centre at Caerphilly provides a point for employers to report incidents which are reportable under the Reporting of Injuries, Diseases and Dangerous Occurrences Regulations 1995, irrespective of whether their activities are controlled by the HSE or local authorities. It provides for the reporting of incidents by telephone, fax, e-mail or hard copy. Use of these arrangements meets the notification requirements of the regulations and reports are passed on to the appropriate enforcing authority. The database held at the centre contains useful information on accident-related issues.

THE MANAGEMENT OF ACCIDENT RISKS

Risk assessments

The Management of Health and Safety at Work Regulations 1999[56] place specific duties on employers to supplement their general duties under the Health and Safety at Work etc. Act 1974 and in particular to undertake assessments of the risks to the health and safety of their employees whilst they are at work.[57] The requirements of those regulations are dealt with more fully in Chapter 3.[58]

---

[55] reg. 13.
[56] S.I. 1999 No. 3242.
[57] reg. 3.
[58] pp.78-84.

Apart from the legal requirement, the following practical reasons for carrying out risk assessments are relevant to local authorities, both as duty holders and enforcers:

1. The absence of risk assessments is a key factor in many of the prosecutions by enforcing authorities.[59]

2. Loss of staff due to ill-health or injury affects the ability to undertake the requirements of the local authority service plan and may increase the pressure on other staff to fill the gap (possibly resulting in work-related stress!). There may be an associated loss of morale.

3. Sick pay costs can be high.

4. Insurance premiums may increase (many local authorities do not insure themselves for all eventualities as it is).

5. Damage to plant and machinery can be expensive, as can the cost of related down-time.

6. The employer may be exposed to the risk of civil claims for compensation. In the case of local authorities, stress has become a feature of such claims.

7. A successful prosecution can result in significant fines and costs, together with the impact on an employer's reputation.

Risk assessment is just the first part of the risk management approach. It provides the information on which judgements can be made about the nature and degree of the risks, and the ways in which the risks should be managed and reviewed. All of the features of a risk assessment are covered in the Approved Code of Practice[60] which accompanies the regulations, but certain key features are worth summarising:

1. The risk assessment should identify all of the risks and hazards associated with work. The level of detail should be proportionate to the risk.

---

[59] See pp.224-226 for examples.
[60] *Management of health and safety at work: Approved Code of Practice* L21, 1999, HMSO.

2. The degree of detail and sophistication will depend on the size, complexity and nature of the hazards present.

3. Assessments should consider all those who might be affected by the undertaking, including members of the public and other workers who may have occasion to visit a site such as sub-contractors, security guards, cleaning and maintenance staff, and include routine and non-routine activities.

4. Steps to identify risks should include examining information relative to the activity involved, including relevant legislation, codes of practice, suppliers' manuals, manufacturers' instructions[61] and information from other competent sources.

5. Risk assessments should be appropriate to the nature of the work and *should identify the period of time for which they are likely to remain valid*. Activities involving frequent changes will require more frequent review.

6. Assessments should not just rely on what is thought or expected to take place but *on what actually occurs*. There is no substitute for observing activities; they often differ from what is supposed to happen. Talking to employees may give a further insight into the actual working practices and they will usually give their opinion as to the risks involved, often relating their experience to "near miss" situations.

7. Existing preventative or precautionary measures should be assessed. If they are inadequate, further steps should be considered to reduce the risk.

For anyone conducting an inspection of premises or activities, whether employer or health and safety officer, it is important that they bear in mind the principles and advice of the Code.

It is also worth considering some relevant court decisions, just to emphasise the importance of risk assessments:

• There need be no more than a foreseeable possibility of injury,

---

[61] They have responsibilities under s.6, Health and Safety at Work etc. Act 1974.

and not a foreseeable probability, for there to be a risk of injury.[62]

- An employer may still be liable for an injury to an employee even though he agrees to an employee temporarily working under the direction of another employer.[63]

- Even a small risk may be foreseeable if a reasonable man would not have disregarded it.[64]

Risk assessment requirements

Enforcement officers need to be aware of the basics of a risk assessment in order to be able to enforce the legal requirements adequately. The Approved Code of Practice[65] which accompanies the regulations states that the assessment should usually involve identifying the *hazards* present in any undertaking, whether arising from the work activities or from other factors such as the building layout, and then evaluating the extent of the *risks* involved, taking account of whatever precautions are already being taken. The Code differentiates between *hazard* and *risk* as follows:

- Hazard is defined as "something with the potential to cause harm".

- Risk "expresses the likelihood that the harm from a particular hazard is realised". The extent of the risk "covers the population which might be affected by a risk, i.e. the number of people who might be exposed and the consequences for them".

The Code summarises the position in that "risk therefore reflects both the likelihood that harm will occur and its severity".

Although there are no fixed rules about how a risk assessment should be carried out, there are a number of guides to help with the

---

[62] *Cullen v. North Lanarkshire Council* 1998 S.L.T. 847 (Inner House). This was a case involving the Manual Handling Operations Regulations 1992, S.I. 1992 No. 2793.

[63] *Nelhams v. Sandells Maintenance Ltd., The Times,* June 15, 1995, CA.

[64] *Gerrard v. Staffordshire Potteries Ltd.* [1995] I.C.R. 502, CA. This involved the Protection of Eyes Regulations 1974, S.I. 1974 No. 1681.

[65] *Management of health and safety at work: Approved Code of Practice* L21, 1999, HMSO. An extremely useful and detailed guide to risk assessment, including worked examples, is contained on the City of Leeds website.

process of risk assessment that reflect the guidance in the Code.[66] These may provide overall guidance on the risk assessment process and/or a more detailed method of risk assessment when dealing with the implications of specific regulations.

There are essentially five steps to a risk assessment, set out below.

Step 1: Look for the hazards
This means looking for hazards which could reasonably be expected to cause harm under the conditions existing in the particular workplace. Such hazards might involve any of the following: fire, electricity, slips and trips, chemicals, working at height, machinery, dust or fumes, odours, transport, manual handling, noise and vibration, inadequate lighting, unreasonable temperatures (high or low), pressure systems, stress.

Before starting the assessment it is important to consider carefully any specific regulations relevant to the working environment and working practices, as they may contain particular requirements.

It is essential to examine the complete working environment and bear in mind that it is not only the employees of the undertaking who may be exposed to risk but also others visiting and using the site. Accordingly, the areas to be examined should include:

• All working areas, including reception and delivery; manufacturing and assembly; stairs, corridors, passageways, lifts and fire escapes; restrooms, washrooms and toilet facilities; service areas such as heating and ventilation facilities; storerooms; accessible roof areas; and activities away from the main place of work such as contract work, home working and sales activities.

• All work activities; machinery and equipment including both routine and non-routine activities, e.g. repair, maintenance and emergency activities; and indirectly related activities such as meal breaks (many accidents have occurred during breaks from work on an employer's premises).

---

[66] *Five steps to risk assessment,* INDG 163 (rev 1), January 2002, HSE. *A guide to risk assessment requirements,* INDG 218, March 2002, HSE.

In identifying the hazards it is useful to talk to employees, examine any accident records and consider any specific literature including manufacturers' guidance.

Step 2: Decide who might be harmed and how
This may include young employees, cleaning staff (employees or contract cleaners), maintenance personnel some of whom may not be on the premises at all times, visitors, customers, members of the public, people sharing the workplace (colleagues or other employer's staff) and equipment operators. Particular consideration should be given to more vulnerable people, e.g. people with disabilities, pregnant women, inexperienced staff and people working alone.

Step 3: Evaluate the risks and decide whether existing precautions are adequate or more should be done
In evaluating the risks it is important to observe working practices and consider whether any defined procedures are being followed. To decide whether more needs to be done it is necessary to consider how likely it is that each hazard could cause harm. Even after taking all precautions there will usually be a residual risk. It is then necessary to decide for each significant hazard whether this remaining risk is high or low. For the hazards identified, it is necessary to ask certain questions:

1. Have all the things required by law been carried out?

2. Have the generally accepted industry standards been adopted?

3. Has everything that is "reasonably practicable" been done?

4. Do the actions taken represent good practice?

5. Is there adequate instruction and training in place?

6. Is the risk reduced as far as possible or can more be done?

Any circumstances giving rise to a serious and unavoidable danger should be recorded and acted upon immediately.

If something needs to be done, it is necessary to draw up an "action list" identifying what is to be done to get rid of the hazard completely or to control it so as to minimise the risk. A time limit should be set, monitored and adhered to. This should be realistic and deal with issues in order of priority. If all of these things have been done then, coupled with the provision of proper information, instructions and training and together with adequate systems and procedures, the risks should be adequately controlled.

All of the consequent safety rules and procedures should be set out and be readily available. In deciding the control measures for each identified risk, it is best to begin by considering what are the most effective measures to take and move down a hierarchical list of measures if the most effective cannot be taken, as follows:

1. *Eliminate the risk* – if possible it is best to eliminate the risk completely, e.g. by using mechanical aids such as trolleys or lifts rather than manual handling.

2. *Substitute the hazard* – use an alternative, e.g. substitute a hazardous cleaning chemical with one which is less toxic.

3. *Contain the risk* – prevent access to the risk, e.g. by the use of barriers or guards.

4. *Reduce the exposure to the risk* – e.g. reduce the time to which a person is exposed to risks such as repetitive strain injury or eye strain by allocating them other tasks at regular intervals.

5. *Training and supervision* – this is a legal requirement in any event but proper training can ensure employees are aware of the risks and the precautionary steps to be taken in conjunction with other measures.

6. *Personal protective equipment* – personal protective equipment must be supplied and used where there are risks that cannot be adequately controlled in other ways, e.g. a noisy environment which can only be reduced to a certain but insufficient degree.

7. *Welfare arrangements* – even when other safety steps have been taken accidents can still occur. In some situations other facilities such as specific washing or medical facilities may be needed.

Step 4: Record the findings
If there are fewer than five employees, it is not necessary to write down the risk assessment results. However, it is good practice to do so and serves as the basis of future assessments. If there are more than five employees, the significant findings of the assessment must be recorded. Employees must also be told about the findings.

An inspector should always ask to see any risk assessment and it will provide him with an indication of the quality of the employer's approach to risk control.

The assessment may quite properly refer to other detailed safety documents to indicate what steps are taken to control risks.

When recording an assessment, it is important to allocate the implementation of any remedial measures to a named individual, to set a realistic timescale reflecting the degree of risk and any interim measures put in place, to introduce any short term temporary measures that may be required, e.g. ceasing an activity, and to monitor progress in achieving the necessary controls.[67]

Step 5: Review the assessment and revise it if necessary
At some point new machinery, materials, substances and procedures will be introduced to most organisations. Any significant changes should result in an assessment to consider any new hazards. Key dates should be set for reviewing the risk assessment. This should be a complete review, checking that the precautions still adequately control each risk. The results should again be recorded. As well as reviewing risk assessments when developments indicate they may no longer be valid, it is good practice to review them regularly at intervals depending on the nature of the risk.

ACCIDENT INVESTIGATION

Reasons for investigation
There are legal and practical reasons for local authorities to investigate accidents at work.

---

[67] Leeds C.C. has included some valuable case studies on its website.

Legal

1. Every local authority has a duty to make adequate arrangements for the enforcement within their area of the relevant statutory provisions, and in accordance with such guidance as the Health and Safety Commission may give them.[68]

2. If local authorities do not investigate accidents in accordance with their duty, theoretically their duties could be transferred to the Health and Safety Executive.[69]

3. There may have been a breach of health and safety legislation by the employer, the employee, the manufacturer and/or supplier of equipment or services, or others such as contractors. This may require enforcement action by prosecution or the service of improvement and/or prohibition notices.

4. The injury may be reportable to the enforcing authority.

Practical

1. Death and injury should be prevented.

2. There may be loss of output and/or work activities disrupted which needs to be resolved as soon as possible.

3. Accident investigation can be used to identify the importance of maintaining health and safety standards to a high level. It can demonstrate the role that employees and management have to play in maintaining safe working conditions.

4. The results of an investigation, especially if used as an element of employee training, can show the impact of inadequate safety performance and encourage employers and employees to play a greater role in accident prevention. The investigation may also reinforce their roles under the Safety Representatives and Safety Committees Regulations 1977.[70]

5. Investigating an accident will enable the investigating inspector to assess the primary and contributory causes, e.g. unsafe

---

[68] Health and Safety at Work etc. Act 1974, s.18(4)(a), (b). See also the mandatory guidance issued under s.18.
[69] *ibid.,* s.45.
[70] S.I. 1977 No. 500.

machinery or plant, the absence of maintenance arrangements, inadequate risk assessments or insufficient employee training.

6. Thorough investigations can be used to assess an employer's arrangements for health and safety management in a specific area connected with the accident, e.g. the operation of fork lift trucks. The investigation might look at the overall planning of their use, maintenance arrangements, employee training, and monitoring and review of the safety arrangements.

7. An accident investigation will identify the overall approach to health and safety and whether a more detailed inspection is required.

8. Enforcement authorities have an obligation to train their staff[71] and accident investigation is an ideal opportunity to develop their knowledge and skills.

9. Whether or not enforcement action is taken, there is also an opportunity to publicise the findings of an investigation in order to increase the awareness of employers, and if necessary the general public, of the risks associated with a particular activity.

10. It is necessary for an enforcing authority to measure its performance in reducing accidents in the workplace against its own planned targets.

Government targets
There is a need to meet the *government's targets* for the reduction in accidents by:

(a) reducing the number of working days lost per 100,000 workers from work-related injury by 30% by 2010;

(b) reducing the incidence rate of fatalities and major injuries by 10% by 2010;

(c) achieving half the improvement under these targets by 2004.

---

[71] Note 6 and Annex 2 to the *Section 18 HSC Guidance to Local Authorities.*

Costs of accidents

The overall cost of accidents to production, in replacement staff, and to society in general are significant. Unfortunately, it is difficult to obtain accurate information on the financial costs of accidents in the local authority enforced sector. However, some indicative costs relating to what are primarily local authority enforced activities give some idea of the significant costs of workplace accidents.

Total cost to employers and society of workplace injury and non-injury accidents and work related illness, by particular industry[72]		
Industry	*Employer costs (£m)*	*Societal costs (£m)*
Distribution and repair	253–900	580–1,310
Consumer/ leisure	138–367	350–610
Hotels/ restaurants	96–333	790–1,050

Local authorities are not immune from the costs of health and safety failings in their own authorities. One study, for example, covered two metropolitan borough council departments employing a total of 1,600 people and looked at health and safety incidents occurring over a 13-week period. The total cost of 45 incidents of injuries and damage was £21,700.

In another case involving a water company, there was a widespread problem of upper limb disorders due to the use of display screen equipment, with 25% of the 650 users experiencing problems. The average cost of each case to the company was calculated to be £5,251. The estimated net savings of introducing a heath care programme over a ten-year period were estimated to be over £885,000.

---

[72] Costs to Britain of workplace accidents and work-related ill health in 1995/96, information supplied in private correspondence by the Health and Safety Executive.

Which accidents should be investigated?

The decision on which accidents to investigate will vary from authority to authority depending to some degree on the resources they have available. Quite often the determining factor will be the seriousness of the incident, for example:

● The severity and scale of actual or potential harm, or the potential for harm associated with the event.

● The seriousness of any potential breach of the law.

● The past performance of the employer in meeting health and safety obligations.

● Any national enforcement priorities (usually issued by the HSC).

● The likely impact of an investigation.

To achieve uniformity of approach, some authorities use a coding system in deciding which accidents to investigate. This is useful provided it is flexible enough to provide for a different approach if the particular circumstances justify it. The coding system used by Birmingham City Council's Environmental and Consumer Services Department is a useful and quite comprehensive example:[73]

The coding system applied to RIDDOR reportable accidents is as follows:

● Code 1: An accident investigation **is** mandatory.

● Code 2: An accident investigation **is** considered necessary.

Note, however, that a Code 2 accident investigation may not be carried out at the judgement of the officer responsible for the premises, given knowledge of premises and past history as recorded on file.

● Code 3: An accident investigation **is not** considered necessary.

Note, however, that a Code 3 accident may be investigated at the

---

[73] Reproduced with its permission.

judgement of the officer responsible for the premises, given knowledge of the premises and past history as recorded on file.

In borderline circumstances, Code 2 accidents should always be investigated.

Guidance on coding
The following will be classified as Code 1:

1. All **fatalities** arising out of work activities.
   (This does not include suicide or death due to natural causes.)

2. The following **injuries:**
   • Amputations (including amputations of digits past the first joint).
   • Permanent loss of sight in one or both eyes.
   • Head injuries involving loss of consciousness.
   • Burns and scolds greater than 10% of the surface area of the body.
   • Any degree of scalping.
   • Crush injuries leading to internal organ damage (e.g. ruptured spleen).
   • Serious multiple fractures (i.e. more than one bone, not including wrist or ankle).
   • Asphyxiations.

3. All of the following types of incident that result in a RIDDOR defined **major** injury:
   • Workplace transport.
   • Electrical incidents.
   • Falls from a height over two metres.

The following incidents (if not falling within Code 1) will be classified as Code 2:

1. All RIDDOR defined major injuries not covered under Code 1.

2. All RIDDOR defined dangerous occurrences.

3. **Any** reportable incident where it appears from the report

that there has been a serious breach of health and safety law (i.e. a notice or prosecution could be required).

4. All cases of reportable diseases, as contained on form F2508A.[74]

5. Incidents which result in an **"over 3 day"** injury, in any of the categories below:
   - Workplace transport.
   - Electrical incidents.
   - Falls from a height over two metres.
   - Incidents which arise out of working in a confined space.

6. Any incident likely to give rise to serious public concern: typical examples are those which involve young children/ persons, vulnerable adults, or multiple casualties.

7. Incidents whose investigation may be relevant to the training needs of a particular officer or student.

8. Those incidents where an investigation is required as part of a local or national project/initiative.

Code 3 covers those reportable incidents that do not fall into the categories of Code 1 or Code 2.

Major injuries and dangerous occurrences
The criteria above rightly take account of the definitions of major injury and dangerous occurrences contained in the Reporting of Injuries, Diseases and Dangerous Occurrences Regulations 1995,[75] as well as the experience of the authority in investigating accidents generally. It is worth setting out some of those definitions as contained in the Schedules to the regulations:

"SCHEDULE 1
MAJOR INJURIES

1. Any fracture, other than to the fingers, thumbs or toes.

2. Any amputation.

---

[74] Approved accident reporting form available from HMSO.
[75] S.I. 1995 No. 3163.

3. Dislocation of the shoulder, hip, knee or spine.

4. Loss of sight (whether temporary or permanent).

5. A chemical or hot metal burn to the eye or any penetrating injury to the eye.

6. Any injury resulting from an electric shock or electrical burn (including any electrical burn caused by arcing or arcing products) leading to unconsciousness or requiring resuscitation or admittance to hospital for more than 24 hours.

7. Any other injury –

   (a) leading to hypothermia, heat-induced illness or to unconsciousness,

   (b) requiring resuscitation, or

   (c) requiring admittance to hospital for more than 24 hours.

8. Loss of consciousness caused by asphyxia or by exposure to a harmful substance or biological agent.

9. Either of the following conditions which result from the absorption of any substance by inhalation, ingestion or through the skin –

   (a) acute illness requiring medical treatment; or

   (b) loss of consciousness.

10. Acute illness which requires medical treatment where there is reason to believe that this resulted from exposure to a biological agent or its toxins or infected material."

Schedule 2 to the regulations defines dangerous occurrences. The list of dangerous occurrences is designed to obtain information primarily about incidents which have a high potential to cause death or serious injury, but which happen infrequently. They are grouped under five headings:

1. *General.* This includes incidents involving lifting machinery, electrical short circuits, breathing apparatus, the collapse of scaffolding, the carriage of dangerous substances by road, the collapse or partial collapse of buildings or structures, certain explosions or fires and the escape of substances in dangerous quantities.

2. *Dangerous occurrences which are reportable in relation to mines.*

3. *Dangerous occurrences which are reportable in relation to quarries.*

4. *Dangerous occurrences which are reportable in respect of relevant transport systems.*

5. *Dangerous occurrences which are reportable in respect of an offshore workplace.*

The dangerous occurrences contained in the first category are those of most concern to local authorities.[76]

A list of 72 reportable occupational diseases is contained in Schedule 3 to the regulations. Most of these will not be found in activities controlled by local authorities, although the following may be more frequently encountered:

- Cramp of the hand or forearm due to repetitive movements.

- Leptospirosis.

- Legionellosis.

- Occupational dermatitis.

The investigation of accidents

In general, speed is of the essence in investigating accidents, particularly where major injuries are involved or there has been a defined dangerous occurrence. This is because:

---

[76] Detailed guidance on the RIDDOR regulations and their interpretation is contained in *A Guide to the Reporting of Injuries, Diseases and Dangerous Occurrences Regulations 1995*, HSE, 2002.

(a) the scene may need to be preserved and examined before any clean-up;

(b) witnesses and any injured people may need to be interviewed as soon as possible;

(c) evidence in the form of photographs, samples and sketches may be required.

However, this is not always possible, especially if an investigation is in response to a written notification rather than an immediate notification by telephone or other direct contact. There will be a number of stages to an investigation:

Notification
This may occur by post, telephone, e-mail or personal visit. Some basic information should be obtained at the time of an initial telephone notification:

● The name, address and telephone number of the person reporting the accident.

● The name, address, telephone number and main activity of the business or undertaking where the accident has occurred.

● The name and address of the injured person and whether they are an employee, visitor to the premises, member of the public, etc.

● The kind of accident and the nature of injuries sustained by the injured person.

● Details of any other organisation notified, e.g. HSE, police, ambulance service.

● Names and addresses of any witnesses known to the person making the notification.

● Whether anything at the scene has been moved or altered as far as they know.

The information obtained during a report by telephone should determine whether the accident is reportable under the Reporting of

Injuries, Diseases and Dangerous Occurrences Regulations 1995, whether the accident is serious and if an immediate investigation is needed.[77] Whoever takes the initial call or receives the written notification must refer the matter immediately to someone with authority to take the decision on the action required. Pro-formas setting out the information required to make the decision are valuable for staff not involved in the enforcement process.

Where possible, the person reporting the accident should be told whether an inspector will be visiting the site and, if so, when; not to disturb the site or remove anything from it; and request witnesses to stay on site or leave their names and addresses if this is impracticable.

Preparing for the investigation
Having decided which accidents need investigating,[78] there are some important advance steps to be taken by an investigating officer, such as:

1. Tell someone where you are going, how long you are likely to be if possible and, if not, tell them how you will report back on your progress.

2. Ensure you take what appears to be appropriate personal protective equipment, e.g. hard hat, suitable goggles, fluorescent jacket and safety shoes.

3. In some cases, especially those that may appear complicated or involve a death, it may be helpful to take along a supporting officer. That officer should be properly authorised and, if you are inexperienced in accident investigation, try to ensure that the accompanying officer is.

4. Remember that it is essential to obtain and capture the evidence at the time of investigation. Afterwards will be too late as the scene will change, as may witness accounts, and evidence may be removed or altered. It may also be necessary to serve notices and take statements at the time.

---

[77] The relevant criteria may include those in HELA LAC 67/1 (rev 2). See also pp.227 and 228.
[78] See pp.245-247.

As investigations may be required at any time, it is valuable to have a standard set of equipment always prepared and ready for use, preferably in a separate bag or case. This should include the following:

- Cameras, perhaps digital and/or video, together with spare films and batteries.

- A variety of sample bags/bag seals/identification tags and labels.

- A notebook (and spare writing implements) and a list of the names, addresses and telephone numbers of useful contacts.

- A dictaphone and spare batteries for situations where it is difficult or impracticable to use a notebook (its contents can be transcribed later and the tape retained).

- Copies of prohibition notices and appeal forms, together with spare carbon papers if required.

- Witness statement forms.

- Section 20 forms.

- Section 19 authorisation forms.

- Seizure/detention forms.

- PACE interview record forms.

- A torch and spare batteries.

- Safety tape/spray paint/hazard tape for cordoning off/marking areas as necessary.

- Marker pens.

- Labels/posters marked "do not touch" or similar.

- Tape measures (short and long).

- A copy of the Health and Safety at Work etc. Act 1974, together with relevant regulations.

It is essential that this bag of equipment is checked and re-stocked regularly. There should be a system for ensuring that this is done as failure to do so can seriously impede an investigation if important items are missing, e.g. camera batteries and statement forms.

The investigating officer should always carry his authorisation under the Health and Safety at Work etc. Act 1974, together with any warrant to enter premises that may be deemed necessary.

Conducting the investigation
If an accident investigation is not carried out thoroughly and promptly, it may form the basis of a complaint to the local government ombudsman. There are few recorded cases, but in a case involving Hinckley and Bosworth DC[79] an injured employee claimed that the council had failed to properly investigate an accident resulting in his serious injury. In particular, he said that the council had failed to keep him informed of progress, delayed making a decision and failed to give him adequate reasons for not prosecuting. It had taken the council almost five months to investigate and decide not to prosecute. The ombudsman criticised the council for its delay in conducting and concluding the investigation and for failing to keep the injured person informed of progress. It ordered the council to make an *ex-gratia* payment of £1,000 to the complainant. The ombudsman concluded that the failings did not result in a wrong decision not to prosecute. Interestingly, however, the complainant expressed a view that the council's failure to interview key witnesses face-to-face affected the adequacy of the investigation and also criticised the council's failure to recognise the importance of CCTV footage.

In conducting an investigation, avoid jumping to conclusions or making assumptions. Facts are what are required and any assumptions not proven by firm evidence are likely to be rejected in any resultant court case or tribunal hearing. The presence of another authorised officer will help to verify the investigating officer's findings, assist in controlling activities on site, ascertain the whereabouts of witnesses and generally assist the orderly

---

[79] Reported in *Environmental Health News*, September 10, 2004.

progress of the investigation. Investigating officers should also be careful to ensure their own personal safety. There is no defined method for investigating incidents but the following should cover most situations:

1. Ascertain any remaining health and safety risks on site and take appropriate precautionary measures.

2. Determine the circumstances of the incident as quickly and as fully as possible. This should include:

    (a) the general site environment;

    (b) any equipment, plant, facilities, working practices or systems involved in the incident;

    (c) the chronological sequence of events leading to the accident.

3. Take photographs of the scene, making sure to include all aspects relevant to the accident, prior to any clean up of the site. Try and control the site clean up to avoid evidence being removed before the on-site investigation is complete, although in some cases the clean up will be dictated by safety considerations. If necessary, the powers of an inspector under section 20 of the Health and Safety at Work etc. Act 1974 may be used. Taking possession of any physical evidence should be done as early as possible. The use of an instant camera may be useful, as will digital and video camera evidence.

4. Taking measurements and drawing sketches will enable scale drawings to be produced and enable the situation to be assessed against any building plans, production, layout and workflow diagrams that may already exist.

5. All witnesses to the events should be identified and listed. These are likely to include eye witnesses, people who may not be able to remain on site for some reason, other employees and managers. It may be desirable to interview them in that order. Interviews should be conducted on site where this is possible and the witnesses are in a fit state to be interviewed. Where this

is not possible, they should be interviewed as soon as practicable and in accordance with good practice.[80] Fully detailed statements should be taken, ensuring that all the facts known to the witness are obtained.

6. Assess the facts and cross check them with the evidence of other witnesses, where necessary, to identify any gaps or discrepancies in the evidence. In such cases consider re-interviewing and making further enquiries to ensure the accuracy and reliability of the information.

7. If conclusions cannot be drawn from the evidence, continue to dig for information until satisfied that the facts enable conclusions to be drawn.

8. Some witnesses will be hostile, especially those with something to hide. If their evidence is contradictory, try and obtain other evidence which allows theirs to be challenged. Always find out as much information as possible.

9. Accidents do not just occur, they are usually the result of a human or mechanical failure. The work practices and systems, and the suitability, age, experience, training and supervision of the people involved should be examined in detail. Safety policies, work instructions, maintenance and repair records, and any employee complaints related to the circumstances leading to the accident should all be examined. Accident records should be scrutinised for indications of any similar incidents.

10. Where specialist help is required to support the investigation, e.g. to take samples of air or substances or to examine plant or equipment, then experts may need to be used, e.g. a chemist or analyst, or an engineer. The Health and Safety Executive has its own in-house experts who will be willing to assist local authorities in appropriate cases. Where in-house expertise is to be called on, it is useful if they have already been authorised under section 19 of the Health and Safety at Work etc. Act 1974. Any experts used, if not already authorised, should be

---

[80] See Chapter 4, pp.174-187.

issued with appropriate authorisation prior to being given access to the investigation site.

11. Ensure that contemporaneous records[81] are kept of all investigative work and interviews. Those records will be essential in the event of any subsequent enforcement action.

12. It is important to keep relevant people informed of action taken and progress made. This is likely to include safety representatives, who may wish to exercise their right to carry out an inspection where there has been a notifiable accident, occurrence or disease,[82] an injured party or their relatives. In some cases there may be evidence of a crime, possibly assault, manslaughter or even murder. In such cases the police must be immediately informed and they may well take over the investigation. Pursuance of a civil claim for damages by an injured party may involve an investigating officer having to give information to that person's lawyer or giving evidence in proceedings in the civil courts. This is something best dealt with through the authority's legal department.

Some very useful information related to the Health and Safety Commission's priority programmes on stress, musculoskeletal disorders, slips and trips, transport safety and falls from height is contained on the HELA training co-ordination site operated by Salford University.[83] This includes information on the nature, investigation and possible control measures for dealing with these matters.

The outcome of an accident investigation

An accident investigation should identify where things have gone wrong, why and who is responsible. It can result in a number of possible outcomes:

1. The need for prosecution.

2. The provision of advice to an employer and/or employee(s).

---

[81] Or records made as soon as practicable after the event.
[82] reg. 6, Safety Representatives and Safety Committees Regulations 1977, S.I. 1977 No. 500.
[83] Details in Biography section of this book.

3. The provision of advice backed up with improvement and/or prohibition notices.

The kind of failings likely to become apparent from an accident investigation might include:

•   The absence of a health and safety policy and safety instructions.

•   Inadequate systems of work.

•   Unsafe plant and equipment.

•   The absence of proper supervision and/or defined responsibilities for health and safety.

•   Insufficient consideration of training needs.

•   Lack of information on health and safety risks associated with particular work practices, equipment and substances.

•   Insufficient employee involvement on health and safety matters.

Most of these issues will occur because of failures to carry out proper safety audits and risk assessments. Part of the enforcing authority's role will be to give advice on these matters as well as taking statutory enforcement action where needed.[84]

Health and safety audits

The absence of health and safety audits is a primary cause of breaches of health and safety law and resultant prosecutions.[85] If an employer, including an enforcement authority, is to meet its obligations under the Health and Safety at Work etc. Act 1974 and the associated regulations, a safety audit is essential and may be regarded as part of the overall risk assessment process. If an enforcement officer is to offer advice, he must be aware of the main elements of the audit, which should include examining the following:

---

[84]   *Section 18 HSC Guidance to Local Authorities*, Annex 2, Statement on Enforcement Policy, HSC, November 2002.

[85]   See pp.224-226 for examples.

Written information

1. The provision of a written safety policy statement reflecting the nature of the activities.

2. Documentation setting out the working arrangements for implementing the policy, including the identification of individual people with specific responsibilities. Job descriptions will need to identify those responsibilities.

3. The ready availability of any internal or Approved Codes of Practice and HSE Guidance Notes relevant to the activities involved.[86]

4. Risk assessments showing the workplace hazards and the safety precautions in place to minimise risk. The risk assessments should include all relevant regulations which impose this requirement.[87]

5. The provision of written safe working systems and safety rules. These should be made available to all relevant employees.

6. Planned cleaning and maintenance systems.

7. The provision of all service and maintenance contracts for plant and equipment together with inspection, service and maintenance records. These should include safety features such as machinery guards, fire protection devices and smoke detectors.

8. Any required test certificates should be readily available.

9. Health and safety training records for each member of staff, demonstrating they have received initial and refresher training in relation to appropriate activities.

10. Records of incidents, dangerous occurrences, diseases and near miss incidents which may not have resulted in injury.

11. Written emergency procedures.

---

[86] In the case of local authority activities this may also include HELA Local Authority Circulars.

[87] See *A Guide to Risk Assessment Requirements*, HSE, INDG 218, March 2002.

12. Procedures for ensuring the health and safety of visitors to the site, including maintenance and other sub-contractors, enforcing authority staff and other visitors.

Health and safety control systems and procedures
These will vary depending on the nature of the activity but the procedures to be put into practice, in addition to being set out in writing, may include:

1. Planned inspection and examination of plant and equipment, storage arrangements, hand tools, hazard warning systems, first aid facilities, personal protective equipment, emergency evacuation plans, etc.

2. Provision of permit to work systems where there is a high degree of risk.

3. The operation of a risk/hazard reporting system.

4. Planned preventive maintenance and damage control systems for plant, vehicles, equipment and the building structure.

5. The provision of first-aid staff, their training and re-training, including training related to any specific injuries likely to arise from particular activities with a relatively high risk of injury.

6. Procedures for monitoring sickness absence and its causes.

7. The practising of fire and evacuation procedures and the regular planned testing of fire alarms and equipment.

8. The provision and monitoring of high quality cleaning schedules which include the removal of hazardous and other waste, spillage clearance and ensuring that all passageways, means of ingress and egress are kept clear.

9. The provision and maintenance of appropriate personal protective equipment.

10. The routine monitoring of the working environment and employees' exposure to it, including temperature, lighting

and ventilation, noise and vibration, and hazardous atmospheres and substances.

11. Monitoring of employees' adherence to health and safety requirements including the correct use of personal protective equipment and personal hygiene.

12. Ensuring that all relevant warning notices and safety signs are provided and maintained.

Training and supervision

Many accidents occur partly as a result of failure to provide adequate instructions, information and training. The arrangements for these should form part of the safety audit. This aspect of the audit should include the following:

1. Ensuring that the manufacturers and suppliers of plant, equipment, materials and substances for use at work have provided proper safety information and advice.

2. Providing relevant and sufficient numbers of safety warnings, instructions and signs, and ensuring they are understood.

3. Ensuring that the individual training needs of staff at all levels in the organisation are regularly assessed and individual training plans are produced.

4. Providing training on induction, changes of job, and when new processes, equipment, legislation or new internal rules and requirements are introduced. Re-training should be undertaken on a planned and regular basis.

5. Ensuring that tasks are only allocated to those who are capable (not merely trained) to do them.

6. General health and safety training should be carried out as well as specific training for potentially skilled or hazardous jobs, e.g. fork lift truck driving, fire fighting or first-aid.

It is frequently suggested that prevention is better than cure and infinitely cheaper. This is something often overlooked, particularly by small to medium sized businesses. It is therefore important that such businesses are encouraged to recognise that health and safety requirements should be taken into account in designing or planning work activities. Budgets, both short and longer term, should contain finance sufficient to meet projected health and safety needs.

Work-related deaths

A work-related death or life-threatening injury are amongst the most distressing situations an investigating officer will have to deal with. It may also involve other enforcement agencies in investigation and prosecution. Accordingly, a protocol has been agreed between the Health and Safety Executive, the Association of Chief Police Officers, the Crown Prosecution Service, the British Transport Police and the Local Government Association.[88]

These organisations have different roles and responsibilities in relation to a work-related death. The HSE and local enforcing authorities, who should also have regard to the protocol, have defined responsibilities under the Health and Safety at Work etc. Act 1974[89] and cannot investigate or prosecute for general criminal offences such as manslaughter. The police and the Crown Prosecution Service will investigate and consider the evidence where a criminal offence such as manslaughter may be involved. Although the protocol makes no specific reference to the role of local authorities as enforcing authorities, they obviously have the same enforcement role in relation to defined activities. Further, related guidance becomes "relevant guidance" for the purposes of section 18 of the Health and Safety at Work etc. Act 1974[90] and therefore becomes mandatory guidance for local authorities.

Protocol principles

The underlying principles of the protocol include the following:

1. An appropriate decision concerning prosecution will be made based on a sound investigation of the circumstances surrounding work-related deaths.

---

[88] "Work-related deaths – a protocol for liaison", February 2003, HSE. For additional guidance, see HELA LAC 22/1, September 2000.

[89] s.18.

[90] "Contact with, and disclosure of information to, the relatives of people killed through work activities", HELA LAC Circular 45/19, April 1992.

2. The police will conduct an investigation where there is an indication of the commission of a serious criminal offence (other than a health and safety offence), and HSE, the local authority or other enforcing authority will investigate health and safety offences. There will usually be a joint investigation but, on the rare occasions where this would not be appropriate, there will still be liaison and co-operation between the investigating parties.

3. The decision to prosecute will be co-ordinated and made without undue delay.

4. The bereaved and witnesses will be suitably informed.

5. The parties to the protocol will maintain effective mechanisms for liaison.

Initial procedure and investigation
The manner in which investigations will be conducted are detailed in the protocol but the key elements specifically related to the enforcing authority involvement can be summarised as follows:

1. In the case of a work-related death[91] or the strong likelihood of death resulting from an incident arising out of or in connection with work, a senior police officer will attend the scene and make an initial assessment. If it is decided that there is a serious criminal offence prosecutable by the police,[92] he will start an investigation, inform the enforcing authority inspector and liaise with him. All decisions should be recorded in writing.

2. The police, HSE and enforcing authority will agree on the use of resources, the disclosure of evidence, interviewing of witnesses and the use of experts.

3. The police and enforcing authority will liaise on arrangements for keeping relatives informed, dealing with the media and making any public announcements.

---

[91] Defined on p. 4 of the protocol.
[92] See paras. 4.1. and 2.1 of the protocol.

4. If the police decide that a charge of manslaughter or other serious offence cannot be justified, the enforcing authority will continue its own investigations, with any agreed police support.

5. Any evidence found during an enforcing authority investigation indicative of a serious criminal offence will be referred to the police without delay.

6. Where there is an investigation, material obtained during the course of the inquiry should be shared, subject to any statutory restrictions on disclosure. Agreement should also be reached on which organisation should be responsible for the retention of exhibits.

Special inquiries
There is provision in the case of some serious incidents, particularly those involving multiple fatalities, for the joint management of investigations.[93]

Prosecution decisions
Any decision to prosecute following a work-related death should be co-ordinated and follow liaison with the police and the Crown Prosecution Service. The prosecution decision[94] should be made known to the accused and bereaved families prior to any public announcements.[95]

Dealing with the bereaved requires a great deal of sensitivity and understanding and it may be the police and/or enforcing authority inspector who undertakes this role. Many of the issues involved are covered in some detail in the relevant HELA Circular.[96]

The decision to prosecute any serious criminal offence (other than a health and safety offence) arising from a death will be taken by the Crown Prosecution Service,[97] with or without any related health and safety offences. The Crown Prosecution Service will

---

[93] Section 6.
[94] See also sections 8 and 9 of the protocol.
[95] Paras. 2.3 and 6.3.
[96] HELA LAC Circular 45/19, April 2002, HSE.
[97] In accordance with the Code for Crown Prosecutors.

relay its decision to the police and the enforcing authorities as soon as practicable so that the authorities can decide whether to prosecute health and safety offences. Where the police and/or the Health and Safety Executive have been involved, the enforcing authority, if it is the prime investigator, will inform them of any enforcement decisions.

Where these various organisations seek to prosecute offences arising from the same investigation, the prosecutions will be initiated and managed jointly. In particular, they should agree on issues such as who will take the lead responsibility; the nature and wording of the charges; arrangements for the retention and disclosure of material; a case management timetable; informing the bereaved and witnesses; announcement of the decision; consultation arrangements; and instructing the prosecuting advocate.

Enforcement action following an accident or dangerous occurrence

The alternative courses of action are described in Chapter 4.[98]

Some further interesting and potentially relevant legal decisions involving accidents are summarised below:

1. Under the Manual Handling Operations Regulations 1992[99] each employer must, so far as is reasonably practicable, avoid the need for his employees to undertake any manual handling operations at work which involve a risk of their being injured. Where that cannot be avoided, he must make an assessment (risk assessment) of those operations and take appropriate steps so as to reduce the risk of injury. In one case, the plaintiff was holding a load of fencing when he caught his footing and fell off a truck. It was held that it was enough for a breach of the regulations if a manual handling operation involved a risk of an employee being injured, irrespective of whether it was to some extent due to the imposition of a load. The court also said that *there need be no more than a foreseeable possibility*

---

[98] p.154. See also Chapter 3, pp.103 and 104.
[99] S.I. 1992 No. 2793.

of injury, and not a foreseeable probability, for there to be a risk of injury.[100]

2. In the case of a long-term employee who had been transferred to light duties due to back trouble, the employee asked to be transferred back to his previous job where he then sustained a back injury whilst lifting a box. It was held that *in view of the length of the employee's service, it was not necessary for the employer to provide repeated instruction to the employer regarding his habits when lifting boxes*. The employer was only under an obligation to provide a safe system of work and to ensure it was followed. In the circumstances, the risk of injury caused by lifting the box was not foreseeable.[101]

3. A refuse collector who was cut by a piece of glass successfully claimed damages for negligence, alleging that his employer should have used wheeled bins for refuse collection. On appeal it was held that, where it was alleged that an employer had failed to provide a safe system of work for his employees, *the court was obliged to consider whether he had taken all reasonable steps to do so*, having regard to the dangers inherent in their duties. In this case the original appeal judge *had failed to take account of the merits and drawbacks of the various alternative systems.*[102]

4. Where a contractor is employed to do work on an employer's premises, the employer must take reasonably practical steps to avoid risks to the contractor's servants which arise from inadequate arrangements made by the employer with the contractor as to how they ought to do the work.[103]

---

[100] *Cullen v. North Lanarkshire Council* 1998, S.L.T. 847 (Inner House).
[101] *Rozario v. Post Office* [1997] PIQR P15, CA. A similar decision was taken in the case of a very experienced window cleaner who was well aware of the risks of his trade.
[102] *Nilsson v. Redditch Borough Council* [1995] PIQR P199, CA.
[103] *R. v. Associated Octel Co. Ltd.* [1996] 4 All E.R. 846.

Chapter 6

PLANNING FOR THE REGULATORY ROLE

THE NEED FOR PLANNING

There are a number of reasons to plan carefully the work required to undertake the responsibility for enforcing health and safety legislation.

The legal duty

Section 18(4) of the Health and Safety at Work etc. Act 1974 states:

"It shall be the duty of every local authority –

(a) *to make adequate arrangements* for the enforcement within their area of the relevant statutory provisions to the extent that they are by any of those provisions or by regulations under subsection (2) above made responsible for their enforcement;[1] and

(b) to perform the duty imposed on them by the preceding paragraph and any other functions conferred on them by any of the relevant statutory provisions *in accordance with such guidance as the Commission may give them.*"

The mandatory guidance referred to in subsection (4)(b) is issued by the Health and Safety Commission.[2] The guidance is intended to act as a framework containing the broad principles local authorities are expected to adopt in enforcing health and safety legislation. It states the elements the Commission considers essential for local authorities to adopt in order to discharge adequately their duties as enforcing authorities.[3] These are:

1. A clear published statement of enforcement policy and practice.[4]

---

[1] i.e. Health and Safety (Enforcing Authority) Regulations 1998, S.I. 1998 No. 494.
[2] *Section 18 HSC Guidance to Local Authorities*, November 2002, HSE.
[3] See s.45 of the Health and Safety at Work etc. Act 1974 for the powers of the Secretary of State to intervene if an authority fails to fulfil its obligations.
[4] See pp.98-102 for more information.

2.   A system for prioritised planned inspection activity according to hazard and risk, and consistent with any advice given by the Health and Safety Executive and Local Authority Enforcement Liaison Committee (HELA).[5]

3.   A service plan detailing the local authority's priorities and its aims and objectives for the enforcement of health and safety.

4.   The capacity to investigate workplace accidents and to respond to complaints by employees and others against allegations of health and safety failures.

5.   Provision of a trained and competent inspectorate.

6.   Arrangements for liaison and co-operation in respect of the Lead Authority Partnership Scheme.

All of these elements are outlined in the section 18 guidance.

Other reasons for planning

There are other reasons for effective planning of regulatory duties, not all of which have a statutory basis, but which are no less important. Overall, more than twelve million people are employed in businesses subject to local authority health and safety enforcement, whilst at the same time there appears to be a declining level of enforcement activity that needs to be addressed.[6]

Controlling work-related illness and injury

The fatal injury rate to employees increased in the years 2002/03,[7] and since 1996/97 there has been an upward trend in the number of fatalities to members of the public, much of the latter being due to increased fatalities in residential care homes.[8] In recent years there has also been an increase in the number of reported major

---

[5]   See pp.32 and 33.

[6]   As indicated in the HELA "Health and Safety Activities Bulletin 2003 – inspection and enforcement in local authority enforced sectors", October 2003, HSE.

[7]   "HELA National Picture 2003 – health and safety in local authority enforced sectors", p. 3, HSC National Statistics, October 2003, HSE. See also statistics produced by the Centre for Corporate Accountability and contained on its internet website.

[8]   *ibid.*

injuries and over-3-day injuries.[9] Over 400,000 people are believed to have suffered illness due to their jobs in the local authority controlled sector, with musculoskeletal disorders and stress, depression or anxiety being the most commonly reported work-related illnesses.[10]

The most recent estimate of working days lost due to injury in the local authority enforced sector is 2,124,000 in 2000/01. Ill health associated with work in the same sector accounted for an estimated 6,100,000 working days lost.[11]

The pattern of injuries varies with the type of work and the nature of the premises in which people work.[12] These factors, together with detailed information based on local knowledge, should contribute to the detailed service plan which each authority is required to produce.

Public and political expectations
Elected members generally show a keen interest in preventing illness and injury at work. This reflects both a natural desire to avoid such problems, as well as their close links with trades unions who also have a vested interest in workforce health and safety. That political interest can quite often be used to help secure the essential resources needed to undertake health and safety enforcement, particularly when competing for scarce resources. Regular reporting on health and safety issues to service committees in relation to prosecutions, accidents, new legislation and changing guidance can influence the views of members on the importance of their authority's role in health and safety.

Local authorities as businesses
Local authorities are large businesses and they have a responsibility to use their manpower and financial resources as effectively and efficiently as possible. This responsibility applies equally to health and safety enforcement, although authorities will have different

---

9   "HELA National Picture 2003 – health and safety in local authority enforced sectors", p. 4.
10 *ibid.*, p. 5.
11 *ibid.*, p. 9.
12 *ibid.*, see the report generally.

views on how this should be achieved, particularly when considered from a corporate viewpoint. However, whatever resources are allocated to health and safety, enforcement powers must be used in the most effective way.

LOCAL AUTHORITY REGULATORY PERFORMANCE

Health and Safety Commission statistics

Some useful indications of the extent of local authority health and safety performance can be derived from statistical returns to the Health and Safety Commission.[13] The latest figures cover the period 2001/02 and are estimates based on local authority returns.[14] A summary of some of the key figures provides a snapshot of local authority activity in 2001/02:

Premises
- Local authorities were responsible for enforcing health and safety in 1,162,000 premises, 3% fewer than the previous year.

- Since 1997/98 there has been a 16% increase (from 185,000 to 215,000) in the number of "other premises" for which local authorities are responsible, e.g. leisure and cultural, consumer services, etc. and a slight drop in other categories.

- The number of wholesale premises has fallen by 16%.

- In the main sectors almost all businesses employ less than 50 staff, with 90% employing less than 10 staff.

Staff resources
- Many inspectors combine health and safety work with other duties, suggesting that perhaps there are relatively few specialist health and safety inspectors.

- There were an estimated 3,480 inspectors holding and using health and safety powers, 4% fewer than the previous year.

---

[13] Published in the HELA "Health and Safety Activity Bulletin 2003 – inspection and enforcement in local authority enforced sectors", October 2003, HSE.
[14] 389 out of 410 authorities submitted details.

- The number of inspectors holding and using section 19 powers[15] fell by 9% between 1998/99 and 2001/02 (from 3,810 to 3,480), while the full-time equivalent fell by 12% (from 1,210 to 1,060) over the same period.

- The number of full-time equivalent inspectors in England has fallen by 15% since 1998/99 whilst the number of premises per full-time equivalent has risen by over 30%. In Wales and Scotland there is a more stable situation.

Visits
- 266,000 visits were made, 11% fewer than the previous year. The number of visits had fallen by 20% since 1998/99, continuing a downward trend in the 1990s.

- Preventive inspections accounted for around 60% of yearly visits.

- Special surveys and activities in 2001/02 accounted for 15,000 visits out of a total of 266,000, a 25% increase on the previous year.

- 19,000 visits in 2001/02 involved accident investigation, 14% fewer than in 1998/99.

- Visit rates per 1,000 premises fell, mainly because of an increase in premises per full-time inspector. As the overall number of local authority enforced premises fell by 3% from the previous year, this may reflect a lower level of inspector resource allocated to health and safety, or greater demand for alternative higher priority work.

- Retail or catering activities accounted for 60% of all visits, although the visit rate fell by 12%.

- Overall visit rates fell by just over 20%, with the rate of visiting only remaining stable in relation to residential accommodation and wholesale activities.

---

[15]   Dealing with the appointment, and specifying the powers, of inspectors.

Notices

The number of improvement notices served has fluctuated slightly over the last few years but generally varies between about 4,700 and 5,000. The same kind of fluctuation is evident with prohibition notices, which total around 1,000 per year. Of more interest in terms of planning enforcement services are the statistics on the premises attracting notices. These may indicate the highest risk areas, which therefore require a greater proportion of available resources:

• The wholesale industry attracts the highest rate/1,000 premises of formal notices.

• In residential accommodation, e.g. hotels, campsites, care homes, the rate of issue of formal notices rose by 60% over the period 2000/01.

• Catering premises attract the highest rate/1,000 premises of informal notices.

• Office-based premises attract the lowest rate/1,000 premises of informal notices.

Whilst statistical information on premises, the level of visiting and degree of formal or informal action will vary between authorities, it can be seen that, in England at least, the level of resources devoted to health and safety enforcement has fallen, whilst the need for services has increased in some sectors of employment. The national statistics certainly give an indication of the main problem areas. Particularly disturbing is the fact that a number of local authorities issued no improvement notices in 2000/01 or 2001/02.[16] Are these cases for the Secretary of State to consider using his powers under section 45 of the Health and Safety at Work etc. Act 1974?

There is no indication of the extent to which local authorities fail to meet inspection targets based on the HELA priority planning criteria. However, the overall fall in inspections and the

---

[16] HELA "Health and Safety Activity Bulletin 2003 – inspection and enforcement in local authority enforced sectors", p. 7, October 2003, HSE.

acknowledged shortfall of health and safety officer resource emphasises the need to target resources at the most risk-prone activities and premises.

THE PLANNING PROCESS

Considerations when producing the plan

It is not the intention here to address all of the elements of business planning; there are plenty of good publications on that process. However, the same kind of formalised approach is required if manpower and financial resources are to be properly used. There are some fundamental things to consider before starting to produce a plan for dealing with health and safety enforcement:

Consider the audience
The plan should consider the needs of those who have to implement it, e.g. enforcement staff. It must address the expectations of elected members who will need to approve it. It must be capable of internal monitoring and external auditing. The plan should be publicly available in the spirit of open government and should be written accordingly.

What should it contain?
It must reflect the mandatory guidance issued by the Health and Safety Commission under section 18 of the Health and Safety at Work etc. Act 1974.[17] The HSC guidance note says that the service plan should include the following:

- Future objectives and major issues that cross service boundaries.

- Key programmes, including a planned inspection programme in the context of the current HSC strategic plan and HELA strategy.

- Information on the services that are being provided.

- The means by which these services are going to be provided.

---

[17]   *Section 18 HSC Guidance to local authorities*, Guidance Note 3, November 2002, HSE.

- Any performance targets and how they will be achieved.

- A review of performance to address any variance from meeting the requirements of the service plan.

It should also contain the health and safety aims and objectives of the authority and the type of resources needed to achieve it. The priority rating of each element of work should be included, together with the manner of monitoring and auditing. It is important to distinguish between statutory requirements and discretionary elements of work. This becomes important if for any reason the plan cannot be achieved, in which case authorities would generally expect to concentrate resources on their statutory obligations rather than those that are merely discretionary. If there are elements of work that are desirable but cannot be achieved, the plan should highlight them. These should be set out in order of priority to be dealt with should resources become available during the plan year.

The language to be used
The best plans are written in plain English and make clear statements about what they propose to achieve and how this is do be done. English may not always be the first language of some of those who may wish to see the plan, so it may be useful for it to be read by others to ensure it can be readily understood.

Who should produce it
The plan will form the basis on which the enforcing authority will undertake its health and safety enforcement work over the plan period. It should, therefore, be prepared largely by, and in consultation with, the staff who have to implement it. They will then support it and be committed to it.

The plan structure
Basically, the plan will define the authority's aims and objectives for its future performance and the ways in which they will be achieved. In practice the process is rather more complicated than this and will usually include:

(a) developing a mission statement for the activity;

(b) deciding on the overall long and shorter term objectives;

(c) describing the means to achieve the objectives;

(d) allocating the resources in the most cost-effective manner.

Most local authority plans will contain at least two elements:

A strategic element
This element will usually contain an overall objective or mission statement reflecting the authority's overall attitude to health and safety, e.g. *"Our mission is to ensure that risks to people's health and safety from work activities are properly controlled. This means reducing risks and protecting people. This covers the health and safety of both workers and the public".*[18]

Whatever the overall statement, *it must be clear and unambiguous and understood by everyone.*

The strategy might also include medium to longer term objectives for health and safety improvement, e.g. the Commission includes in its strategic plan goals which are included in the *Revitalising Health and Safety* Strategy Statement[19] although local authorities often include such detail in their service plan aims and objectives.

Service aims and objectives
The plan should state what the service intends to do, the resources to be allocated to individual tasks and may include reference to the way in which performance against the plan will be monitored.

The *aims* might include:

• The protection and promotion of the health, safety and welfare of those affected by work activities, whether at work, in the home or elsewhere.

• The enforcement of the Heath and Safety at Work etc. Act 1974 and the relevant statutory provisions at those premises allocated to the authority for enforcement purposes.

---

[18] Taken from the Health and Safety Commission Strategic Plan 2001-2004, October 2001, HSE.
[19] *ibid.*, pp. 8, 9.

- To provide advice and assistance to employers and the general public on health and safety issues.

- To provide a health and safety service consistent with the council's corporate aims, objectives and strategies.

- To ensure public confidence that the health, safety and welfare of employees and the public is regularly monitored and that employers adopt the highest standards.

- The regular review of health and safety activity in order to respond to changing circumstances and ensure continuous improvement.

These aims have been taken at random from published service plans. Local authorities are not uniform in their plans; the important thing is that they best reflect their attitude and approach to health and safety.[20]

Certain principles may also be considered in developing the way in which authorities plan and carry out their work. The following are taken from the Health and Safety Commission's Strategic Plan 2001-2004. Some of them may already be included in corporate and/or service plans. If not, they may act as a useful guide in the development of service plans:

- Work in close collaboration with stakeholders, looking actively to increase their engagement and promote full participation in improving health and safety.

- Be transparent and open about what we do, why and how, sharing what we know with others.

- Work to develop new relationships with those stakeholders who we have not traditionally reached, to secure their engagement and participation in improving health and safety.

- Promote partnerships both between the enforcing authorities and stakeholders and between employees and employers to develop and disseminate good practices which improve standards of health and safety.

---

[20] Examples can be found on most local authority websites.

- Pay particular attention to the needs of small firms and vulnerable groups, including ethnic minorities, women and people with disabilities in developing our programmes.

- Ensure that our actions are consistent, proportionate, targeted and transparent.[21]

- Take action on the basis of sound evidence about health and safety problems and the costs and benefits involved.

- Monitor and evaluate our programmes to learn from what we can find, ensure value for money and assist in the benchmarking of future programmes.

Service *objectives* should say what it is intended to do in each element of the service, how the aims will be achieved and lead in to a statement of the resources to be allocated to that area of activity. Some of the objectives may include:

1. *Routine inspections.* The establishment and maintenance of planned inspection programmes based on the guidance on inspection programmes and an inspection rating system issued by HELA.[22] Following this advice will help local authorities to demonstrate they are making "adequate arrangements" for enforcing the law as required by section 18 of the Health and Safety at Work etc. Act 1974.

2. *Investigation of complaints.* To investigate all complaints alleging unsatisfactory working conditions or matters associated with working practices likely to have an adverse effect on the health, safety and welfare of employees and/or others.

3. *Accidents.* To investigate accidents involving fatalities, major injuries and dangerous occurrences[23] and reportable diseases in accordance with any guidance issued by the Health and Safety Commission and HELA.

---

21 As defined in the HSC section 18 guidance.
22 Described in HELA LAC 67/1 (rev 2), December 2000.
23 As defined in the Reporting of Injuries, Diseases and Dangerous Occurrences Regulations 1995, S.I. 1995 No. 3163.

The criteria for deciding which accidents to investigate may be set out in more detail in the authority's Enforcement Policy Statement.

4. *External liaison.* To liaise with other organisations having an involvement with and/or interest in health and safety at work, including the Health and Safety Executive, Trading Standards Authority, the Fire Authority and other government and related agencies.

    As part of its strategy for 2001-2004, HELA has identified some areas of activity for joint working between the Health and Safety Executive and local authorities[24] which should result in:

    (a) inspection of areas of significant risk which previously have not been inspected;

    (b) planned programmes of inspection in work areas where both the Executive and local authorities work;

    (c) more sharing of knowledge and experience between HSE inspectors and local authority enforcement officers; and

    (d) increased use of HSE's specialist resources in areas where local authorities have no expertise.

    There is some indication that certain authorities may not want to be involved in joint working if they have not been involved in the initial planning of this work, but it should be possible for the respective organisations to sort out this difficulty.

5. *Lead authority partnership scheme.* To act as the lead authority (in relation to any established scheme in the authority) and/or to participate fully in respect of any other externally led scheme.

6. *Health and safety education.* To provide targeted health and safety training to employer organisations and to participate where practicable in other relevant training schemes. To act as

---

[24] HELA LAC 40/6, March 2003.

a point of contact for those employers wishing to establish employee training programmes.

Most local authority enforced businesses have fewer than ten staff and limited resources for conducting training activities. Local authority initiatives can be targeted at priority activities known to be a problem in their area or directed to those priority issues identified by the HSE, such as slips, trips and falls. Setting this type of objective also allows for involvement in the training activities of specific training organisations such as trades unions, Chambers of Commerce, local colleges, engineering employers' federations, etc.

7. *Staff training/competence.* To provide and maintain a sufficient number of fully trained and competent staff commensurate with their duties and responsibilities, and to provide personal development opportunities consistent with their abilities and expertise.

It is important to have sufficient numbers of suitably trained staff to meet the requirements of the service plan. Staff should be allocated to duties consistent with their training, skills and level of authorisation under section 19 of the Health and Safety at Work etc. Act 1974. It is wasteful of manpower, and expensive, to allocate highly experienced inspectors to low priority inspection work or to use relatively inexperienced staff on complex accident investigations, although in the latter case they may gain useful experience when assisting more experienced colleagues.

8. *Special projects.*[25] Dedicated inspections and training will be undertaken where activities or premises have been identified nationally and/or locally as giving rise to a particularly high level of risk. This enables the authority to highlight in its resource allocation priority issues such as stress, slips, trips and falls, together with the higher risks in places such as warehouses and residential homes. Some useful examples of special projects include:

---

[25]   Apart from the stated examples there are many others on the HELA training co-ordination website.

(a) *warehouse safety group*[26] – following an invitation to warehouse businesses to identify their health and safety needs, a joint local authority/business group was established to provide information and training to employers. This has now become a self-help group run by employers who share information and advice, with the local authority providing advice, secretarial support and training in association with equipment manufacturers. As a result of greater employer awareness, resource has been released to develop similar initiatives involving care homes and leisure services;

(b) *small business pack*[27] – a health and safety advice pack together with a self-audit checklist has been found to be a valuable aid to employers, as well as other local authorities keen to follow this initiative;

(c) *health and safety matters for small firms*[28] – the development of a service to help small firms secure safe and healthy workplaces includes a health and safety information system which is also available on-line, relevant workshops and seminars, regular newsletters and an advice line;

(d) *on-line risk assessment*[29] – an extremely useful risk assessment package for small and medium sized businesses providing detailed information on what the process involves, including worked examples.

9. *Auditing.* The management auditing of the health and safety enforcement service will be undertaken at intervals of not less than x years and an action plan produced to ensure service improvement. This may involve internal and external auditing. This should ensure continuous review and improvement.

Authorities may well choose alternative objectives and wording; that is not important. There are no set rules for what should be

---

[26] Basingstoke and Deane D.C.
[27] *ibid.*
[28] Nuneaton and Bedworth B.C.
[29] Leeds C.C. This won a HELA award in 2003.

included in a statement of aims and objectives. There should, however, be a proper indication of what the authority expects and proposes to do. Unrealistic expectations should not be set as they will almost certainly fail, leaving staff disillusioned and morale low.

Resource allocation

The amount of staff resource allocated to the service programme tasks should be based on good quality management information supplemented by well-informed judgements where detailed information is not available. The collection of relevant data to assist the planning process should be ongoing.

Deciding how much manpower resource to allocate to individual tasks is not a precise science but it is important to try and get some indication of the amount of time required. This can be done by some basic monitoring. It is useful to keep records of the amount of time spent on key tasks for a given period, say for three months, which can be followed up at a later date to check the accuracy of the data or any assumptions made. They key tasks might include:

- Primary inspections of premises.

- Follow-up visits to check compliance.

- Accident investigations.

- Court proceedings.

- Complaint investigation and requests for advice/assistance.

- Education and training activities (including the associated monitoring).

- Auditing activities.

- Involvement in local authority corporate work.

- Attendance at special working party meetings.

- Committee meetings.

- Special projects.

- Administrative work of field officers.

- Monitoring different methods of working.

It is arguable that the greatest contribution to health and safety is achieved by field officers enforcing legal requirements in the workplace and, therefore, the more time spent on that activity the greater the potential benefits to the health and safety of employees. Monitoring of this kind should not only identify what resources are used on these and other tasks, but also reveal where time savings and alternative methods of work may improve an inspector's performance. It also provides an opportunity to compare individual performance and examine the reasons for any differences.

It can be quite revealing to establish the amount of time spent on administrative duties. Examining the way these are performed and the available options can result in alternative, time-saving methods of work. Looking at work patterns can also be useful, for example: operating some kind of shift pattern; concentrating field work into one longer period outside the office rather than returning for lunch; working a series of longer days; concentrating on specific seasonal activities at certain times of the year; working from home; and "blitzing" some types of work activities to create maximum local impact and publicity.

It may be useful to use similar topic headings to those indicated above when considering the actual allocation of resources.

Routine inspections
Authorisations of inspectors to undertake health and safety functions should reflect their skills and experience. It is vital not to allocate tasks to those who do not have the necessary skills or knowledge, as they will struggle and fail to meet expectations if tasks are allocated on an individual basis. If work is allocated on a team/ division basis it enables the group to take responsibility for performance and to allocate individual tasks between them in whatever way they may choose. It is important to use individual skills carefully. Asking an experienced officer to perform relatively minor tasks may not be the best use of his skills and certainly will not be the most cost-effective.

Guidance on the formulation and implementation of inspection programmes is contained in HELA circular LAC 67/1 (rev 2).[30] Adoption of this guidance will help demonstrate that "adequate arrangements" are being made for health and safety enforcement. The circular sets out the main components that should be contained in an inspection programme.[31] Authorities will decide the allocation of total resources to each component. The circular emphasises the need to ensure a regular programme of visits to the highest hazard/risk premises and the allocation of sufficient resources to investigate accidents and complaints. The circular also contains details of the HELA inspection rating system which is designed to help authorities prioritise and decide inspection frequencies.[32] The rating system relies on risk assessment and the numerical weighting of a number of specific factors: safety hazard; health hazard; safety risk; health risk; welfare; public risk; and confidence in management. The method of assessment and rating is set out in detail.[33]

Whilst many authorities may wish to devise and use their own inspection report forms, the Health and Safety Executive's Field Operations Directorate has published an inspection report form which may also be of interest.[34]

Investigation of complaints/accidents
The investigation of some complex complaints or accidents may involve more than one officer, with the most experienced taking the lead, but a supporting role in investigations will also add to the skills and knowledge of the other officer(s). These different levels of resources should be identified in any allocation of resource in the plan.

External liaison
Important as involvement with other health and safety related bodies can be, allocating operational staff to these bodies if they are non-productive should be avoided. Something that is merely a "talking shop" is not worth the effort.

---

[30]   December 2000.
[31]   Annex 1.
[32]   Annex 2.
[33]   Appendices 3-5.
[34]   Details contained in HELA LAC 22/14, June 2001.

Health and safety education
Well-directed education can have significant benefits if targeted at the areas of greatest risk. It should be geared not only to increasing the knowledge of the participants but also to improving the level of self-regulation in the activities or premises involved. It may be argued that the local authority role should be mainly as a facilitator rather than provider of such services because the resource demands are too great. Where officers are involved in educational activities, they need to be well trained to deliver this work and must also wish to undertake it. If not, use them elsewhere.

Staff training/competence
Well trained and competent staff are essential to fulfilling the requirements of any service plan. Good quality training will include both group training, e.g. in response to new legislation, and individual training in response to specific individual needs. This is discussed in more detail elsewhere.[35]

Special projects
There is always scope for introducing special projects into a work plan. They provide a departure from the routine of standard forms of inspection whilst introducing new and potentially challenging activities. They will vary according to local circumstances and any HSC priorities. Therefore, in addition to the Commission's priority of addressing slips, trips and falls, etc., local authorities have been seen to give priority to such matters as:

- Risk assessments of playground equipment at catering and leisure facilities.

- Training sessions for staff at residential care homes to reduce the incidence of back injuries.

- Education and training of warehouse staff to minimise the incidence of injuries due to falls, manual handling and transport movement.

- Development of risk assessment and manual handling advice for inclusion on a local authority website.

---

[35] See pp.293-302.

Auditing

The mandatory section 18 guidance states that "To ensure that any management system is effective, it should be frequently monitored, reviewed and audited. A quality audit may be defined as a systematic and independent examination to determine whether quality activities and related results comply with planned arrangements and whether these arrangements are implemented effectively and are suitable to achieve specified objectives."[36] More details on the auditing process and its possible results are contained elsewhere.[37]

Format of the service plan

There is no right or wrong way of setting out the plan. There is often a "house style" in use in authorities, but it is useful to keep the key features of each element of the plan together, for example:

1. Objectives and sub-objectives.

2. Work activity, e.g. premises or activity inspection.

3. The priority of each type of activity, i.e. high, medium or low.

4. Resources to be used on each activity, e.g. senior inspector, inspector, technical assistant, etc., and the amount of time allocated, e.g. in man-hours or days. Allowance should be made for annual leave, anticipated sickness levels and any predicted staff movements.

5. The time-scale for completion of the tasks. This may vary from an all year round activity to a specific time period to reflect seasonal issues or a dedicated period for a promotional or special initiative.

6. Auditing and review periods in advance of the preparation of the following year's service plan.

The associated costs will be reflected in the relevant annual budget.

Reviewing the plan

Authorities may take different views on this essential process but the key factor is the timeliness of the review process. It is no good

---

36 Guidance Note No. 4, October 2002.
37 See pp.286-293.

leaving the review until the end of the year, as this is too late to make any changes if things have not worked out as planned. Indeed, without regular reviews it will not be known whether the plan is on target.

Monthly monitoring and reviews of performance will provide a snapshot of activity and show if work is basically on target and groups or individuals are performing to expectations. If any individual targets are not being met, the reasons can be addressed at that time. Quarterly reviews will give a better indication of how the overall plan is being met and may be a time to make fundamental changes if the plan is off course, e.g. as a result of less resource than expected, a higher level of accidents or complaints than anticipated, a greater amount of time spent on special projects than planned, new priorities or budget changes. This may also be a time to review the possible impact of unpredicted events, e.g. new legislation or a change in political expectations for the service. Plans should be flexible enough to accommodate change whilst trying to adhere to statutory responsibilities.

Reviewing the service plan should involve the staff responsible for implementing it. They are often best placed to know the reasons for any variances and should share the responsibility for, and commitment to, any changes required.

Ongoing service plan reviews enable proactive responses to be made, rather than the knee-jerk reactions that follow from a sudden awareness that all is not well.

Built into any service plan should be time not only to review performance against the plan but also to address other important issues, for example:

(a) alternative and potentially more effective ways of working;

(b) the use of customer surveys to assess the response of employers, employees, trades unions, complainants, etc. to the service provided;

(c) updating the premises/activity database to ensure resources can be directed at the highest risk areas;

(d) making contact with authorities with reputations for good enforcement practice to learn from their experience;

(e) maintaining contact with the Health and Safety Executive's Local Authority Unit to ensure awareness of the latest health and safety developments.

AUDITING THE MANAGEMENT OF HEALTH AND SAFETY ENFORCEMENT

The section 18 mandatory guidance advises[38] that appropriate techniques should be used to measure performance against agreed standards and benchmarks, and to ensure that policies are being adhered to and that the aims and objectives of the organisation are being achieved, with the results and recommendations arising from any review process such as benchmarking, peer review and auditing being developed into an action plan for continuous improvement. It is also expected that all enforcing authorities should have in place arrangements to promote consistency in the exercise of discretion, including liaison arrangements with other local authorities.

The Health and Safety Commission has worked with HELA to produce an inter-authority auditing protocol[39] which provides a framework to help local authorities assure themselves that their performance is consistent with the Commission's Strategic Plan and the section 18 guidance. It is accompanied by a HELA circular.[40] It is expected that auditing will help local authorities to identify and share good practice, compare their performance to ensure consistency, identify areas for improvement, promote good enforcement practice and secure continuous improvement on service delivery. Local authorities are expected to undergo an audit of their enforcement practice at least once every five years and these audits can also be part of any Best Value reviews.

---

[38] Guidance Note No. 4, October 2002.
[39] "Auditing framework for local authorities' management of health and safety enforcement", HELA, June 2003.
[40] HELA LAC 23/19, January 2002. See also HELA LAC 23/20, July 2003 for guidance on calculating the section 18 compliance levels.

Auditing framework

The HELA auditing framework is broken down into a number of key components intended to address defined criteria, as follows:

1. *Enforcement policy and procedures.* The local authority should have a published enforcement policy in compliance with the HSC's section 18 guidance.

2. *Managed work programme.* The local authority should have adequate arrangements in place to ensure that all health and safety enforcement is targeted, proportional, consistent and transparent[41] and have a system for monitoring that this is so.

3. *Training and competence.* The local authority should have adequate arrangements in place to ensure that all health and safety enforcement is carried out by competent officers in accordance with the HSC's guidance.

   The local authority should establish a framework in which it may liaise and co-ordinate with Local Authority Associations, the Health and Safety Executive and professional bodies, to ensure that enforcement officers are properly trained and competent.

4. *Investigation of accidents, complaints about health and safety conditions (requests for service) and complaints about the local authority.* Adequate arrangements should be in place to respond to accidents (also incidents of ill-health, dangerous occurrences, etc.), complaints about the local authority and complaints about workplace health and safety conditions.

5. *Review and quality assessment of the local authority's management of health and safety enforcement and development of action plans.*

Requirement to undergo audit and develop an action plan

The auditing framework deals with the section 18 mandatory requirements and good practice. It may need revision to include other matters if an authority wishes to use the framework but

---

[41] See explanation in HSC section 18 guidance.

extend it to other aspects of its health and safety activities. The audit questionnaire contains a series of relevant questions, space for the auditors' comments and a considerable number of auditors' notes containing questions and comments to assist in ensuring that all significant matters are taken into account. It forms a very useful basis for a comprehensive audit.

The audit protocol was designed originally to be used as a desk-top exercise within inter-authority groups or external auditors, but visits to check policies and procedures in practice are, it is suggested, essential for a meaningful audit.

London Inter-authority Health and Safety Audit[42]

An extremely valuable and informative exercise in inter-authority auditing has been conducted under the auspices of the Association of London Environmental Health Managers. Whilst its findings may not be used to draw conclusions about health and safety performance in other UK authorities, many of the issues raised, problems found and examples of good practice may be experienced by other local authorities. In particular, the good practices found may well be worth considering by other local authorities facing difficulty in meeting the HSC's section 18 expectations.

The audit was carried out to assess the London Boroughs' compliance with the section 18 guidance issued by the Health and Safety Commission. It involved 33 of the 35 London Boroughs, with the remainder being audited by the Health and Safety Executive's Local Authority Unit. The audit process was designed to complement the government's Best Value regime and the achievement of best practice. Its other aims included:

• To help protect public health by ensuring effective enforcement of health and safety law.

• To maintain and improve consumer confidence.

• To identify and disseminate good practice to aid consistency.

• To provide information to inform health and safety policy.

---

[42]   Summary report, June 2003.

- To allow the local authority to comply with section 18 of the Health and Safety at Work etc. Act 1974.

- To provide the means of identifying poor performance requiring local authority management intervention.

- To identify common challenges that could be better dealt with on a regional or national strategic level.

The audit assessed each local authority's conformance with the section 18 guidance, using the HELA audit protocol.[43] It looked at five areas:

1. Enforcement policy and procedures.

2. Managed work programmes.

3. Competence and training.

4. Investigation of accidents, requests for service and complaints.

5. Review and quality assessment.

The careful planning, selection and appointment of highly qualified practitioners, their training by HSE Local Authority Unit staff and the structured approach to the exercise give the results considerable credibility.[44] This is also a good example of the benefits of joint working between the HSE and local authorities, as reflected in the HELA Strategic Plan for 2001-2004.[45]

Main findings
These can be summarised under particular topic headings set out below.

Resources
Insufficient staff was cited as the main reason for not meeting planned programmes of work.

---

[43] HELA LAC 23/19.
[44] The audit process is dealt with in detail in section 2 of the report.
[45] See also HELA LAC 40/6 for more detail on joint working proposals.

Enforcement policy and processes

There were examples of well written, clear, comprehensive and well-monitored policies. Some policies were generic and not specifically related to health and safety. Some required updating or review. Where there were omissions these included:

• Keeping a public register of notices.

• Actions on permissioned work, e.g. asbestos removal.

• How the authority would implement in practice transparency, targeting and consistency.

• Lead authorities.

• Competency requirements.

• Disclosure of information to employees and employee representatives.

• Defining responsibility.

• Proportionality.

• Accountability.

• Regulatory approaches from advice to prosecution.

Other issues included consultation on, and publication of, policy; communication of the inspection outcomes; inadequate filing systems; inadequate procedures relating to health and safety; ineffective management of health and safety work; limited systems for ensuring consistency; and inadequate co-ordination of work which crossed team boundaries. There were inconsistencies in the use of inspection pro-forma, notice checking, the use of letters, etc.

Management of work programmes

There were a few cases where service plans did not address health and safety, or plans did not adequately deal with the following:

• Plan reviews.

• The resources required.

- Educational and promotional activities.

- The HSC priorities of slips, trips and falls, workplace transport, etc.

- The methods to achieve the declared inspection programmes.

- Lead Authority Partnerships.

Other issues identified included the inaccuracy of some databases which made forward planning a hit and miss affair; the absence of internet access; variances in risk rating of premises; and the absence of any planned inspection programme. Common issues relating to the service of notices concerned the proper identification of accompanying schedules; matters relating to appeals; and the use of the most recent HELA guidance.

Training and competence
There were good examples of measures taken to ensure consistency of enforcement, but issues arose in the case of some authorities that included the following:

- The competency assessments needed updating together with documented evidence to that effect.

- A lack of procedures for the authorisation of officers and the need to review some that did exist.

- Training needs assessment and monitoring were not always present, although some were reviewed annually.

Investigations
Some authorities did not have documented policies containing the criteria used for investigating accidents and fatalities. There was also some inconsistency between food and health and safety teams (a common problem in other areas of health and safety work).

In some cases there were no documented policies for dealing with customer complaints or service requests.

Performance monitoring and review
This was one of the areas where shortcomings were especially evident, including:

- The absence of systems to measure, monitor and review performance against performance indicators contained in service plans or enforcement policies.

- Many databases were inaccurate.

- The use of peer review exercises or benchmarking was lacking.

- Some authorities had no arrangements for customer feedback.

Examples of good practice

The inter-authority audit did what was intended and identified shortcomings in the approach of some authorities which are now being addressed. It also identified some very good practices. Equally important was the willingness to share the results of the audit with other authorities who may gain from the auditing experience and findings. The audit also identified some excellent practices/guides which other authorities may well find it useful to consider. A list of these is appended to the report but includes:

1. *Monitoring service quality* – a protocol for monitoring health and safety risk assessed inspections, records and paperwork, outstanding statutory notices and the monthly monitoring of performance targets.[46]

2. *Service of notices* – a procedure for ensuring a consistent approach to the service of notices. It involves a checking system to ensure that all of the required information is included in a notice prior to it being served and a follow-up system for checking compliance with notices.[47]

3. *Guidance on the preparation of prosecution reports* – a procedure for formatting and submitting prosecution reports, including key information required.[48]

4. *Guidance on the issue of formal cautions* – information and advice on the circumstances in which a formal caution is appropriate and the procedure to be adopted.[49]

---

[46] Royal Borough of Kensington and Chelsea Environmental Health Department.
[47] *ibid.*
[48] London Borough of Wandsworth Environmental Services Division.
[49] *ibid.*

5. *Accident, including fatal accident, investigations* – detailed criteria, including out-of-hours arrangements, and practical and administrative procedures for investigating accidents.[50]

6. *Monitoring the nature, consistency and quality of safety and licensing inspections* – a procedure which sets out the type of activities to be monitored, the frequency of monitoring, the method of recording information and the options for corrective action.[51]

7. *Induction and competency* – a framework for assessing the professional competency of health and safety staff new to the department. This is linked to an individual work and training plan which is carefully monitored.[52]

STAFF TRAINING

There does not appear to be a well co-ordinated and consistent strategy or approach to training undertaken by local authorities, although the Health and Safety Commission has produced a policy statement on health and safety training.[53] However, the section 18 mandatory guidance includes, amongst the requirements that local authorities have to meet to discharge adequately their duties as enforcing authorities:

"provision of a trained and competent inspectorate".[54]

Annex 2 to the guidance summarises the units and essential elements to be addressed in providing the appropriate competences for both Health and Safety Executive and local authority health and safety inspectors.

The legal position

The requirement to provide adequate training applies both to duty holders in the local authority enforced sector and to local authorities themselves. The Health and Safety at Work etc. Act

---

[50] London Borough of Hammersmith and Fulham Environment Department.
[51] *ibid.*
[52] *ibid.*
[53] Contained in HELA LAC 84/3 rev, July 2000.
[54] *Section 18 HSC Guidance to Local Authorities*, p. 1, November 2002, HSE.

1974 places a duty on the employer to provide "… such information, instruction, training and supervision as is necessary to ensure, so far as is reasonably practicable, the health and safety at work of his employees."[55] The Management of Health and Safety at Work Regulations 1999 provide that every employer shall provide his employees with "comprehensible and relevant" information[56] on the risks, control measures and safety procedures to be used.

In particular, an employer has a duty to provide employees with adequate health and safety training[57] when they first start work for that employer, when they are transferred to any new role or given a change of responsibility, etc. The training has to be repeated periodically where appropriate.

These requirements can presumably be taken to require a local authority to ensure that health and safety officers have been adequately trained to deal with the kind of risks they may expect to encounter. This is an important and potentially onerous responsibility and the Approved Code of Practice which accompanies the regulations makes it clear that, when allocating work to employees, employers should take account of the employees' ability to carry out the work without risk to themselves or others. Accordingly, if a local authority enforcement officer is, for example, sent to investigate an accident, but has not been properly trained in how to assess risks or carry out an investigation and suffers an injury as a result, the authority may be liable under health and safety legislation and may also be exposed to a civil claim for damages.

Although there are few cases dealing with any actions against local authorities on this issue, certain things may be deduced from decisions made in some relevant cases, for example:

• An employer's training programme must be designed for the individual employee's needs.[58]

---

[55] s.2(2)(c).
[56] reg. 10(1).
[57] *ibid.,* reg. 13.
[58] *Warner v. Huntingdon D.C.* (2002) L.T.L. 16 May, 2002, CA.

- No matter how experienced the employee, health and safety information and training must still be provided.[59]

- No training programme can cover all eventualities but an employer cannot just organise basic training. He must consider the less common but foreseeable activities carrying a risk of injury.[60]

Training of local authority enforcement officers has to be an ongoing process. It has to be continually reviewed and modified to take account of new developments, e.g. new enforcement responsibilities and changing conditions.

Strategy for health and safety training

Most local authorities appear to have a training policy but not all seem to have a structured approach. It may be helpful for authorities to consider the Commission's strategy as a guide to the development of their own training arrangements.

The HSC strategy seeks to:

(a) bring about a substantial improvement in the quality and quantity of health and safety training;

(b) raise awareness of the importance of health and safety training;

(c) promote an awareness of the importance of competence in controlling risk; and

(d) influence providers of the education system to provide the necessary framework of basic knowledge and skills.

The HSC states that "the strategy is directed at engaging HSE and local authority enforcement officers in assessment of the competence of workers and managers and the adequacy of training by employers. This assessment forms an important part of inspection, investigation and enforcement activities." This is fine

---

[59] *Jeffrey Russell O'Neil v. D.S.G. Retail Limited* [2002] E.W.C.A. 1139 Civ, CA, July 31, 2002.

[60] *Philip Michael Chalk v. Devises Reclamation Co. Ltd., The Times*, April 2, 1999, CA.

so long as local authority officers have received adequate training in the first place. If not, their own competence becomes suspect as does their ability to carry out enforcement duties consistent with HSC guidelines.

The strategy also suggests that local authorities should:

(a) prepare their own, or adopt the Health and Safety Commission's, strategy on health and safety training;

(b) implement that strategy;

(c) assess the adequacy of training provided by employers as part of inspection, investigation and enforcement;

(d) promote health and safety training in local media, following successful enforcement action on training issues;

(e) support and work with others who provide health and safety training.

Although most service plans appear to contain an element of external or in-house training for employers on specific aspects of health and safety, e.g. slips and trips and warehouse safety, there appear to be relatively few training strategies aimed at employees. However, local authorities are both enforcers and duty holders and should at least have developed training strategies for their health and safety enforcement staff in order to meet their section 18 obligations.[61] The findings of a study on local authority health and safety enforcement training produced for the Chartered Institute of Environmental Health in 1997 make depressing reading.[62] The overall findings can be summarised as follows:

1. Overall training within Environmental Health Departments was conducted in a fragmentary manner, lacking co-ordination and integration.

---

[61] Staff training may also be subject to the Health and Safety (Training for Employment) Regulations 1990, S.I. 1990 No. 1380. See also HELA LAC 84/2, September 2000.

[62] "Health and Safety Enforcement – competency, consistency and training", Chartered Institute of Environmental Health, 1997.

2. Too much reliance was placed on casual and unfocussed induction procedures and "discredited" cascade training methods.

3. There were, nevertheless, some good training initiatives being undertaken through regional health and safety liaison groups.

4. Many authorities had in place structured and well-managed programmes for identifying health and safety training needs, although training evaluation varied between authorities.

5. Many authorities still had much to do in addressing training issues.

There is no quantitative information on how much local authority training has developed and improved in the intervening period.

Identifying training needs

Local authority personnel departments will generally have responsibility for developing and overseeing corporate training initiatives but it is important that the specific needs for health and safety officer training are addressed by the employing department. More particularly, a named individual should have responsibility for developing and co-ordinating the departmental strategy and training plan. It should be a core element of the service plan. An extremely useful guide on local authority health and safety enforcement training has been produced by HELA.[63]

Individual needs
An appropriate officer, usually the line manager, should identify the training needs of the individuals under his control. This process may include:

●   Deciding the experience and qualifications needed to undertake their defined role.

●   Assessing their individual development needs to allow them to progress to a higher level within the organisation if they so wish.

---

[63]   *Managing training for health and safety enforcement officers – a guide for first line managers*, HELA. The work was undertaken by the Department of Environmental Management, University of Salford.

- Evaluating their field performance to see if it meets departmental expectations.

- Assessing their ability to enforce health and safety legislation in accordance with the enforcement policy and section 18 guidance.

- Considering their ability to deal with complex issues.

- Deciding on their ability to manage their own personal workload and work within a team environment to achieve departmental and/or corporate objectives and targets.

- Assessing personal development needs.

- Personal or professional development interviews with a senior manager are an essential part of the process of evaluating individual needs. Adequate time should be allowed for these. Although many interviews will be on a one-to-one basis, there may be cases where the individual being interviewed prefers the presence of another colleague (usually of senior status). The most likely reason for this would be if they felt intimidated by the interviewer or considered the interviewer would be adversely biased in his judgements.

A number of local authorities include in their quality procedures the detailed arrangements for ensuring the training and competency of their health and safety officers. Ideally, these procedures will identify the basic qualifications and knowledge expected, induction and monitoring arrangements, training evaluation and monitoring and supervisory arrangements.[64]

Divisional/team needs
In addition to any individual training needs there will also be training requirements that cover a wider group of people. The topics involved may include:

- New legislation, codes of practice, circulars and EU directives.

---

[64]   A useful example is contained in the Quality Procedures manual of Nuneaton and Bedworth BC Environmental Services Department.

- New policies and procedures, whether committee, departmental or corporate.

- Evaluating and implementing change arising from customer surveys.

- Service plan implementation and review.

- New or additional responsibilities, e.g. new activities or authorisations.

- Addressing any non-conformity issues identified from service audits.

- Introduction of new equipment or work procedures.

- Revised methods of working.

- Continuing professional development requirements.

The training needs and the estimated resources in time and manpower to achieve them should be fed into the service plan proposals. The training needs should be prioritised to fit in with the overall strategy and service plan. The plan finally agreed should be set out in tabular form identifying the individual, the training to be undertaken and the time periods for implementation. It will usually contain elements of in-house training, any corporate training courses, and external and post entry training. Any staff training needs not approved should be notified, with reasons, to the individuals involved.

Implementation and monitoring of the plan
Producing the plan is the relatively easy part. Implementation requires careful planning to ensure not only the right quality of training but the availability of good quality trainers at locations that are conducive to the type of training being undertaken. The Chartered Institute of Environmental Health runs a number of specific courses designed to meet the needs of local authority enforcement officers, as do the Health and Safety Executive and some local education colleges.

Once the plan has been agreed and prioritised, it is important to monitor the quality of the training and the individuals and organisations that deliver it, through evaluation. Equally important is the ongoing assessment of the individuals undergoing training to ensure that they benefit from it. Feedback should be encouraged from those undertaking training and those delivering it to ensure it remains up-to-date and relevant. Feedback is the most important part of any evaluation process; without it the exercise is pointless.

The training plan should be part of the overall service plan. As such, its preparation, development, implementation and monitoring should follow a similar cycle.

Competences for health and safety and local authority inspectors

The following contains the summary of the units and essential elements in the standards for Occupational Health and Safety Regulation published by the Employers' National Training Organisation and contained in the HSC's section 18 guidance.[65] There are a number of universities and other training organisations, including the Chartered Institute of Environmental Health, able to provide training courses covering these competences. This list provides a reminder to any local authority reviewing its training activities.

Unit R1: Identify the plans and priorities of the regulatory authority for work-related health and safety, and contribute to them effectively
- R1.1 – Identify the objectives, plans and priorities of the regulatory authority for work-related health and safety and personally contribute to them effectively.

- R1.2 – Manage time effectively to ensure the efficient use of resources.

Unit R2: Inspect duty holders, worksites and activities for the purposes of work-related health and safety regulation
- R2.2 – Prepare for inspections of workplace health and safety for the purposes of regulation.

- R2.2 – Conduct inspections of workplace health and safety for the purposes of regulation.

---
[65] Annex 2.

- R2.3 – Report on the conduct and findings from inspections of workplace health and safety for the purposes of regulation.

Unit R3: Investigate work-related accidents, incidents, ill-health reports and complaints for the purposes of health and safety regulation
- R3.1 – Prepare for investigations of work-related accidents, incidents, cases of ill health or complaints for regulatory purposes.

- R3.2 – Determine immediate action needed to ensure effective investigation and manage continuing risk.

- R3.4 – Evaluate the extent of intervention and enforcement needed for regulatory purposes.

- R3.5 – Manage and conclude investigations.

Unit R4: Plan and gather evidence for the purposes of work-related health and safety regulation
- R4.1 – Plan the taking of evidence for the purpose of work-related health and safety regulation.

- R4.2 – Gather and preserve evidence for the purpose of work-related health and safety regulation.

Unit R5: Enforce statutory provisions and brief a prosecutor for the purposes of work-related health and safety regulation
- R5.1 – Prepare reports recommending prosecution for alleged breaches of work-related health and safety legislation.

- R5.2 – Initiate and report on prosecution proceedings.

Unit R6: Enforce statutory provisions and present guilty pleas in a magistrates' court for the purposes of work-related health and safety regulation
- R6.1 – Prepare reports recommending prosecution for alleged breaches of work-related health and safety legislation.

- R6.2 – Initiate legal proceedings for alleged breaches of work-related health and safety legislation and present the prosecution case in a magistrates' court, when a guilty plea is entered by the defendant.

Unit R7: Draft and serve notices or other statutory enforceable documents for the purposes of work-related health and safety regulation
- R7.1 – Draft and serve notices or other statutory enforceable documents for the purposes of work-related health and safety regulation.

Unit R8: Influence health and safety duty holders and others for the purposes of work-related health and safety regulation
- R8.1 – Work with duty holders and others to establish work-related health, safety and welfare standards, procedures and management arrangements in force in the organisation.

- R8.2 – Influence duty holders and others to improve work-related health, safety and welfare standards, procedures and policies.

- R8.3 – Communicate externally with duty holders, employee representatives and external parties.

- R8.4 – Communicate internally with colleagues.

Unit R9: Improve work-related health and safety through promotional activities
- R9.1 – Plan and contribute to local projects and initiatives to promote work-related health and safety.

- R9.2 – Promote work-related health and safety awareness through dissemination of appropriate information.

Not all of these competences are covered by every health and safety training programme but organisations such as the Chartered Institute of Environmental Health and Salford University can provide information on what training arrangements are available to fulfil these criteria.

ALTERNATIVE AND/OR INNOVATIVE WAYS OF WORKING

Many authorities find themselves short of inspectors and need to consider alternative ways of working to make the best use of scarce resources. Even if fully staffed, it is useful to raise the profile of health and safety by the use of alternative or innovative ways of working. It should also be remembered that health and safety was intended to be self-regulatory. It is, therefore, useful to consider ways in which employers can be encouraged to adopt a more responsible and self-regulatory approach to managing health and safety, rather than relying on the enforcing authorities to police their activities. There are many proactive and innovative authorities willing to develop different ways of dealing with health and safety issues. Some of these are a response to resource issues, HSC initiatives or quite simply a desire to secure best value for money. Examples of these different approaches to health and safety enforcement are given below. Anyone interested in following up some of these initiatives and good practices should contact the health and safety enforcement departments (usually environmental health services) of the authorities direct.

Keeping track of businesses and priority inspection areas

This is a familiar problem for enforcement authorities. Leeds CC used a technical officer to find and identify premises with priority hazards, thereby providing an improved property database and identifying activities requiring higher priority action. The authority also developed an online self-assessment tool dealing with hazards and risks.

Targeting priority activities

Some authorities are targeting priority areas in different ways. These may include the HSE's priorities and/or those specific to the individual local authority area. They include the following:

1. Guildford BC has devised accident investigation scoring criteria which weight accidents according to a number of factors. This includes the kind of criteria recommended by HELA as part of its advice on priority planning.[66] It allocates scores to particular features of the incident, with scores above a defined figure automatically triggering an investigation. Accidents with scores between other parameters are investigated at an officer's discretion in discussion with his senior officer.

---

[66] HELA LAC 67/1, December 2000.

2. Birmingham CC has adopted a similar approach using a coding system applied to accidents reportable under the Reporting of Injuries, Diseases and Dangerous Occurrences Regulations 1995. The details are contained in the Environmental Services Department accident investigation policy. Systems of this kind aid the better targeting of resources, yet allow officers to use their discretion depending on particular circumstances.

3. Bradford MDC operates a "high hazard" programme of inspections, with some officers concentrating exclusively on specific risk areas such as call centres, window cleaning, steel stockholders and warehouse transport.

4. Halton BC and a number of other authorities have specifically addressed key topic areas, including slips, trips and falls as the main components of their routine planned inspections. Whilst other obvious health and safety issues are not ignored, this approach ensures that the main perceived risks to occupational health and safety are addressed.

5. Selby DC has used questionnaires to obtain health and safety information relating to its lowest risk premises, releasing resources to concentrate on inspecting a backlog of higher risk activities. Completion of the higher risk activity inspections revealed that many were incorrectly rated. The more accurate picture of risk will no doubt allow a more realistic assessment of service needs.

6. Taunton Deane BC targeted catering premises, sending out advance information to premises to be visited, and then structured inspections around selected risks.

7. Christchurch BC targeted warehouses with fork lift trucks, sending out self-inspection questionnaires and a newsletter on workplace transport safety.

8. A particularly innovative and successful scheme to help small and medium-sized businesses has been established by Nuneaton and Bedworth BC. Designed to inform businesses on health and safety issues and legal requirements, the initiative

comprises a health and safety information system, workshops/ seminars and training, publication of newsletters and an advice line.[67] Its dedicated website has averaged over 8,100 "hits" per month since its launch in September 2001 and, at the time of writing, exceeds 11,000 per month.[68]

Employee training and education

A lot of authorities devote time to training on specific high risk issues:

1. Telford and Wrekin DC and Chichester DC have run health and safety seminars for staff of residential care homes.

2. Tower Hamlets LBC and other authorities have conducted seminars on slips, trips and falls.

3. St. Albans City and DC produced a "Workplace Transport Sample Risk Assessment" booklet to help employers in this area of activity identify and control the related hazards.

4. Caerphilly CBC organised a health and safety seminar on workplace transport, supplemented by relevant safety information packs.

5. St. Albans City and DC produce a number of "self help" booklets, the most recent being guidance in relation to the prevention of upper limb disorders.

6. A lot of authorities appear to have organised seminars on the management of workplace stress.

7. Authorities have also organised seminars on matters such as legionella, asbestos, violence and risk assessment.

---

[67] Details in "Health and Safety Matters – a small firms initiative in support of revitalising health and safety", Steve Moore, paper to CIEH East Midlands Centre Conference, April 2003.

[68] The initiative received the National HELA Award for Innovation 2000.

BIBLIOGRAPHY

A Guide to Risk Assessment Requirements, INDG 218, March 2002, HSE.

A Guide to the Reporting of Injuries, Diseases and Dangerous Occurrences Regulations 1995, November 2002, HSE.

A Practical Guide to Criminal Investigations for Local Government Officers, Andy Bowles, Chadwick House Group Ltd.

A strategy for workplace health and safety in Great Britain to 2010 and beyond, February 2004, HSE.

Auditing framework for local authorities' management of health and safety enforcement, June 2003, HSE.

Enforcement, Better Regulation Task Force, April 1999.

Enforcement Concordat, March 1998, Cabinet Office.

Enforcement Guide (England and Wales), HSE (regularly updated).

Five steps to risk assessment, INDG 163 (rev 1), January 2002, HSE.

Freedom from fear, Union of Shop, Distributive and Allied Workers.

Health and Safety Commission *Business Plan 2002-2003,* June 2002, HSE.

Health and Safety Commission *Business Plan 2003-2004,* July 2003, HSE.

Health and Safety Commission *Enforcement Policy Statement,* January 2002, HSE.

Health and Safety Commission: Pilots to explore the effectiveness of workers' safety advisors – interim findings, March 2003, York Consulting.

Health and Safety Commission *Strategic Plan 2001-2004 – summary,* September 2001, HSE.

Health and Safety Commission *Strategic Plan 2001-2004,* October 2001, HSE.

Health and Safety Enforcement – competence, consistency and training, 1997, Chartered Institute of Environmental Health.

Health and Safety in the UK, Jeremy Stranks, 1998, Pitman Publishing.

Bibliography

"Health and Safety Matters", *Environmental Health Journal,* August 2003, Chartered Institute of Environmental Health.

Health and Safety Policy 2000, Chartered Institute of Environmental Health.

Health and Safety Training – what you need to know, October 2001, HSE.

HELA *Annual Report 2002,* November 2002, HSE.

HELA *Health and Safety Activity Bulletin 2003,* October 2003, HSE.

HELA *National Picture 2002,* November 2002, HSE.

HELA *National Picture 2003,*October 2003, HSE.

Lead Authority Partnership Scheme – liaison on health and safety in the local authority enforced sector, November 2001, HSE.

London Inter-Authority Health and Safety Audit – Summary Report, June 2003, Association of London Environmental Health Managers.

Managing Training for Health and Safety Enforcement Officers – a guide for first line managers, HELA.

Memorandum of Understanding, 2001, HSE.

National Picture 2002: Health and safety in local authority enforced sectors, October 2003, HSE.

Need Help on Health and Safety? Guidance for employers on when and how to get advice on health and safety, January 2002, HSE.

Occupational Health and Safety Law, Brenda Barrett and Richard Howells, 1997, Pitman Publishing.

Passport Schemes for Health, Safety and the Environment – a good practice guide, October 2003, HSE.

Planning and Decision Making, Jon Sutherland and Diane Canwell, 1997, Pitman Publishing.

Procedures for enforcing health and safety requirements in Crown bodies, June 2001, Cabinet Office.

RIDDOR information for doctors, October 1996, HSE.

Safety Lottery – how the level of enforcement of health and safety depends on where you work, September 2003, UNISON.

Section 18 – Health and Safety Commission guidance to local authorities, November 2002, HSE.

Stone's Justices Manual, annual, Butterworths.

Successful health and safety management, December 2000, HSE.

The Code for Crown Prosecutors, 2000, Crown Prosecution Service.

The enforcement of regulatory offences, Claire Andrews, Sweet and Maxwell.

The Health and Safety Executive – working with employers, November 2000, HSE.

Understanding health surveillance at work – an introduction for employers, November 2001, HSE.

Voices from the frontline – a report on shopworkers' experience of work-related violence and abuse, Union of Shop, Distribution and Allied Workers.

Work-related deaths: a protocol for liaison, February 2003, HSE.

Circulars

Advice to local authorities on inspection programmes and an inspection rating system, HELA LAC 67/1 (rev 2), December 2000, HSE.

Application of the HSW Act sections 3 and 4 and HSE's arrangements for enforcement, Operational Circular OC 130/3, January 2002, HSE.

Choice of appropriate enforcement procedure, HELA LAC 22/1, September 2000, HSE.

Contact with, and disclosure of information to, the relatives of people killed through work activities, HELA LAC 45/19, April 2002, HSE.

Contact with representatives and employees at visits and disclosure of information, HELA LAC 73/2 (rev), November 2000, HSE.

Enforcement Management Model – general guidance on application to health risks, Operational Circular OC 130/5, July 2002, HSE.

Enforcement Management Model (EMM) – introduction and training arrangements for local authorities, HELA LAC 22/18, July 2003, HSE.

Guidance for other people to accompany enforcement officers on site, HELA LAC 22/2, September 2000, HSE.

Health and Safety Commission strategy on health and safety training, HELA LAC 84/3 (rev), July 2000, HSE.

Health and Safety (Enforcing Authority) Regulations 1998, HELA LAC 23/4 (rev 2), December 2000, HSE.

Health and Safety (Enforcing Authority) Regulations 1998: A-Z guide to allocation, HELA LAC 23/15, January 2000, HSE.

HELA protocol for inter-authority auditing of local authorities' management of health and safety enforcement – guidelines for local authorities, HELA LAC 23/19, January 2002, HSE.

HELA Strategic Plan 2001-2004: Developing joint working between LA's and HSE in 2003/04, HELA LAC 40/6 March 2003, HSE.

Inspection report forms, HELA LAC 22/14, June 2001, HSE.

Investigation of complaints, HELA LAC 22/5, July 2000, HSE.

Lead Authority Partnership Scheme, HELA LAC 44/3, September 2000, HSE.

LA enforcement in premises in which they may have an interest, HELA LAC 22/10, April 2000, HSE.

Liaison between HSE and local authorities, HELA LAC 48/2 (rev 2), October 2000, HSE.

Major incident response procedures, HELA LAC 20/2, March 2000, HSE.

Prosecuting individuals, Operational Circular OC 130/8, July 2003, HSE.

Sentencing in criminal proceedings – aggravating and mitigating factors, the Friskies and Howe judgements, Operational Circular OC 178/2, January 2002, HSE.

Serious incidents involving agency workers, HELA LAC 2/4, September 2000, HSE.

The cautioning of offenders, Circ. 18/1994, Home Office.

The Health and Safety (Enforcing Authority) Regulations 1989: HSWA section 6 enforcement, HELA LAC 23/5, November 2000, HSE.

The Health and Safety (Training for Employment) Regulations 1990, HELA LAC 84/2, September 2000, HSE.

The HELA health and safety indicator of performance: guidance on calculating the section 18 HSWA compliance levels, HELA LAC 23/20, July 2003, HSE.

USEFUL WEBSITES

The following sites contain useful information on health and safety related topics particularly relevant to local authorities.

Many individual local authority websites contain details of enforcement policies, work plans, advice on specific health and safety topics, details of local initiatives, etc. and therefore are a further source of useful information.

www.trainingco-ord.org – this is the HELA training co-ordination website managed by Salford University. It contains information on the HSE Priority Programmes; examples of local authority good practice, campaigns, initiatives, guidance documents and legal case studies; a directory of local authority liaison groups; and health and safety training information.

www.hse.gov.uk/lau– contains current HELA LAC circulars; annual reports; newsletters; audit protocols; statistics; plans and strategies; press releases; guidance on enforcement issues and other health and safety related publications.

www.hse.gov.uk/enforce – full details of the Health and Safety Enforcement Guide.

www.cieh.org – information on the Chartered Institute of Environmental Health training courses; health and safety publications; library information; and responses to consultation documents.

www.cabinetoffice.gov.uk – details of the Better Regulation Taskforce and Enforcement Concordat.

www.dca.gov.uk – Department for Constitutional Affairs site dealing with human rights issues; general advice on the Human Rights Act; information on the civil procedure rules; civil procedure practice directions and criminal appeal rules.

www.employmentappeals.gov.uk – employment appeals procedure, past judgements on human rights and contracts of employment (little on health and safety – information best found in law libraries or legal texts).

www.lacors.com – website of the Local Authorities Co-ordinators of Regulatory Services. It provides information on local authority activities across a wide range of services including health and safety, promotes quality regulation, and provides and disseminates comprehensive advice, guidance and good practice information.

INDEX

320   *Index*